# How Rockets Fly

## STUDENT'S EDITION

### Math & Science Learning Standards Applied to Rocket Design Grades 4-6

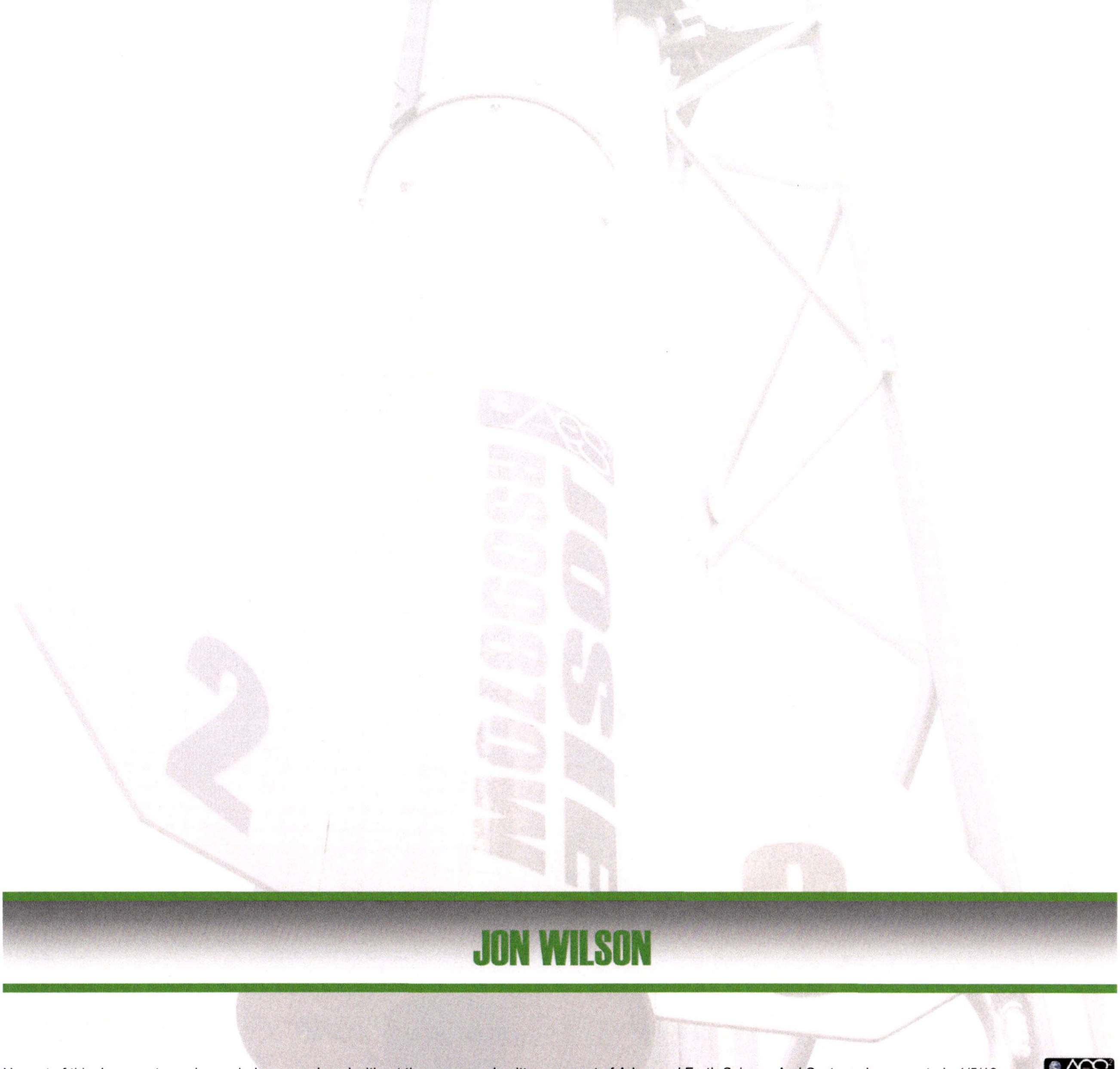

JON WILSON

# PREFACE

# PREFACE

In 1998, I was asked to arrange to have some aerial photographs taken of a manufacturing site. They needed some pictures of the site for an engineering feasibility study. The need at the time was urgent, but when I tried to arrange to have aerial photographs taken, I was informed that there was a six week wait. Due to the high cost of flying and maintaining an airplane, a long list of customers was first required to warrant a flight.

A colleague of mine who was aware of my love of rocketry asked if I could build a rocket to take the photographs. We agreed that a rocket had the advantage of mobility, and could deliver daily, up to date aerial photographs. We also agreed that rockets would cost much less than owning and operating an airplane. At the time, there were few complex UAVs (Unmanned Aerial Vehicles) or drones. To avoid having a long delay like this occur in the future, work on a design to take photographs from a rocket began.

When I first started building rockets as a child, I had the awful experience of watching my rockets gracefully drift out of sight, no doubt swallowed up by some huge maple tree or landing gracefully on someone's roof. It was because of those experiences, the idea of using a "jumper's canopy" or *parasail* to recover a rocket was born. This would help us control the descent of the rocket and minimize the risk of losing it. In addition to the parasail, a "pop out" nosecone stabilizer was necessary to change the center of pressure during the descent stage for this rocket. So an entirely new special canopy system design was developed to control the descent stage of the flight.

After nine months of working evenings and weekends, a rocket was readied for its first test launch and certification. Even though the test launch was somewhat a success, my choice of materials, the weight of the rocket, and the ability to disassemble and service the rocket in the field left much to be desired. So I began to work on the next generation design.

We also wanted to integrate a video camera into the design to transmit live video during the flight. This meant anyone miles away could tune a TV to channel 59, and watch the launch as though they were in the rocket. This turned out to be a very useful analysis tool for making comparisons between the flight simulations and the actual flight performance. Moreover, while viewing a video monitor on the ground, pilots could experience the flight as though they were actually inside the rocket. And during the descent, having eyes from the rocket would make it much easier to aim the camera and steer the rocket back to the launch site.

So the new design incorporated a digital still camera, a video camera with video transmitter and transmitting antenna, seven control servo motors and their receiver, a parasail, a stabilizer, an emergency parachute, and an electronics bay for power and circuitry requirements. **With the rocket fully loaded, including fuel, it weighed in at just ten earth pounds.** Only three pocket tools were required for complete field assembly and disassembly.

As my children witnessed their crazy Dad working feverishly to overcome the obstacles of the improved design, they thought it would be a great idea to launch the rockets at schools. After several test flights, I decided to give it a try. The reaction from the children, faculty, and news media was beyond my expectations. Even so, something was missing. I wanted to do more than just launch a rocket. I wanted to share more about the development of a rocket and demonstrate how they came to be. I needed to get organized. I needed a course outline to demonstrate the process.

# PREFACE

During this time I was also teaching at a local technical college in the engineering department, and had stumbled over a rather surprising discovery. While strongly encouraging the students (more like demanding the students if you were to ask them) to write very concise lab reports using a very systematic approach, it became apparent they were using and retaining abundantly more of the course topics. Since they were writing technical documents reflecting what they had learned, they began to notice dramatically improved technical reading skills.

Why not teach these concepts to younger children and give them the opportunity to use these tools throughout their education? Rockets, especially the rockets used in this book combine all of the sciences: mathematics, physics, mechanics, electronics, chemistry, and yes, even art.

The combination of this discovery, along with the lure of rocketry, seemed ideally suited for each other. So I began to write this book. Shortly after I began to write the labs for this book, another friend suggested that I integrate the learning outcomes, now referred to as the learning standards. Running with yet another suggestion from a friend, *How Rockets Fly* came to be. The first copy of this book was written for the classroom in 2003 but was never published. So in 2018 the book was readied for online publication.

The labs contained in this book are written to build a very strong foundation in problem solving. The step by step process in each of the labs intentionally shows each step; from the idea, to the hands-on application of that idea. The steps followed in this book are the same steps required for solving any problem, simple or complex, and best of all, they work every time. In my experience, as a teacher and in the engineering consulting field, drawing a picture of the problem and then labeling the picture is often overlooked. As such, this book focuses on first drawing a picture of the problem, then labeling the picture. Once these first two critical steps are accomplished, the remaining steps required for solving a problem become much clearer.

In order to support the original labs in this book, the Mobile Launch Trailer, MLT, was built to house all of the support equipment needed for a launch, and for the collection of data required to complete the labs. A computerized launch sequencer to automate the launch, complete with all the safety bells and whistles, and live video monitoring of each flight; all operated remotely from the MLT.

As you can probably imagine, there have been times during this whole process when I began to feel like all of this was a crazy idea, and that I too might be crazy. On a few occasions, I thought the book might be more appropriately titled, ***How Rockets Fly…or Don't***. But I always managed to pull myself together by just remembering all of this really wasn't my idea. I'm just executing the ideas of others. I reflect back on all the friends and their suggestions. Its then I realize…I'm not crazy; they are. That means you Rick, Samantha Marie, Phil, and Jan.

So get ready to have some fun with science. I can only hope that this learning experience will be a favorite of both teachers and students, and something they will cherish for a lifetime. I know I will.

Jon Wilson

# TABLE OF CONTENTS

# STUDENT'S GUIDE TO *HOW ROCKETS FLY*

# STUDENT'S GUIDE TO *HOW ROCKETS FLY*

4$^{th}$ - 6$^{th}$ Grade Learning Standards Applied to High Powered Rocket Flight and Descent

## Table of Contents



## List of Figures

## 1 Introduction

Hello there! I hope you're as excited about reading this book as I am that you're reading it. The *Students Guide* you are reading right now is a brief explanation of how the rest of this book is organized and why.

You're going to learn skills here that will empower you and help you understand HOW to read math and science books more effectively. You will know what to look for and how to grasp new topics quickly, which in turn will free up more of your time for other wonderful learning experiences in history, the arts, and languages, just to name a few.

Best of all, this book is for everyone! If you're a person that enjoys math and science, this book will help get the most out of your time as you pour through all of the exciting topics in math and science. If you're like me, you're probably astounded by everything surrounding you and you're probably continually asking yourself questions like "how does that do that?" or "why does that do that?" I have always found math and science a great way to help me understand so much more about my surroundings.

Even so, it's not everyone's cup of tea. So if you're an artist, a musician, or someone that would like to spend less time belaboring over math and science, and more time with the piano or saxophone, this book is definitely for you too! And by the way, you would not believe how many interesting physics properties there are to musical instruments. Even sheet music has very interesting properties relating to the binary number system.

The most important thing I want to say here is that this book is not supposed to be treated as an assignment, but as a guide for you to use from this point forward as you go through your journey of education.

All of the following labs reinforce the topics already being taught in the classroom by your teacher but apply these topics to the exciting world of rocket science.

As you progress through the labs, you will gain more and more understanding of problem solving and rocket science, and you'll have plenty of questions too!

I encourage you to read the material and note in the book where you have questions. Remember that asking questions is how we all learn when we get stuck.

If there are parts of the book you don't feel comfortable about, don't be shy about asking your teacher. That's what teachers are for--they are here to help you get through the rough patches in your education.

Remember, questions are always good if you've read the material and attempted the work. But more often than not, you will still find you may need more explanation; you'll have more questions. Just remember to give it your best shot first, and then ask your teacher for help.

The rest of this guide will get you familiar with each of the sections and sub sections of the labs and will explain why the labs are formatted as they are. All of the labs follow the same format. So let's get started!

## 2 Format of the Labs

This book is divided up into four (4) Labs, and each Lab is divided up into seven (7) sections. The rest of this guide will explain a little bit about the sections in each lab. Each of the labs builds on the previous Labs, but you'll find the format of each lab identical.

As you read this guide, you may find it useful to refer to one of the Labs that follow.

The *Student's Guide to How Rockets Fly* introduces you to the organization of the labs. ***See Figure 1***.

1. Abstract
2. Nomenclature (glossary of terms)
3. Introduction and Purpose
4. Procedure
5. Results
6. Conclusions
7. References

**Figure 1: The Seven Sections to each Lab**

There are seven steps in each of the labs. If you look at ***Figure 1***, you'll see each step listed. In this guide, we'll take a closer look at each of these steps, from the abstract to the references sections, so you can get a better understanding of why we're doing each step.

In ***Figure 1***, there's a step called the "Procedure". When we get to that step, we need to divide it out into several steps just to make sure we can retrace and document the steps we used to find our solutions.

So Step 4 of the Procedures in ***Figure 1*** is actually broken up into to seven smaller steps. You can see these all listed in ***Figure 2***.

1. Draw a picture of the problem
2. Label the picture
3. Write a mathematical expression that defines the picture
4. Solve for the unknowns
5. Calculate the results (by hand, by calculator, or by computer application software)
6. Create a lab to verify or deny the work in 1-5
7. Test the results of 1-5 in the lab

**Figure 2: The Seven Steps to problem solving**

# 3 About the Format

In a typical learning standard lab, the *abstract, will state the learning standard, a mathematical law, or a hypothesis*. The remainder of the lab will *prove*, *apply* and *test* the learning standard.

At first, the abstract or learning standards may seem a little strange. Since it typically introduces a new topic, you may not fully understand it until the lab has been completed. **Don't Panic!**

Once you complete the lab, you should go back and read the abstract once more. The unfamiliar topic first introduced in the abstract will probably make a lot more sense to you.

This will be true for each lab. The goal of each of these labs is to progress from an abstract concept to a comprehensive understanding of the concept.

Your teacher may decide that some portions of the lab may be discussed briefly, or even omitted, while other sections embellished. It is up to the teacher to decide how much of the lab is actually discussed or completed based on the available time. Of course, you can always do as much or even more on your own.

**Be sure to share what you learn with others!**

# 4 The Format in Detail

## 4.1 Abstract:

*The abstract defines some law, principal, or learning standard*. One or more learning standards are introduced for you and applied in each of the labs. The applications of the learning standards in this book focus on the use of these principals as they apply to the design and flight of rockets.

It is common for the word "abstract" to sound a bit strange to most people. It's a word not often used. Standing alone and without full explanation, the contents of an abstract are typically not understood.

But after the lab is completed, the reader can re-read the abstract and likely find it much easier to understand and apply its meaning.

*There's a thing called abstract art. When you first look at abstract art, it may be difficult to understand what the artist is trying to portray. But give an artist the time to explain its meaning and then the art can be fully appreciated.*

### 4.2 Nomenclature (glossary of terms):

*The nomenclature (another weird word) section of each lab defines terms* that will be used in the lab. The definitions come after the abstract, and before the rest of the lab. This is done so that you will be familiar with the new terms used in the remainder of the lab. The nomenclature section provides the reader with a quick reference before, and during the reading of the lab. These terms are conveniently assembled at the beginning of each lab for quick reference.

### 4.3 Introduction and purpose:

The introduction and purpose portion of each lab *will explain the reason why the topic is important, and give you an application of the concept*. Here, you will find the reason why the lab is being performed and how it can be applied to a real life application; in our case, rockets.

The solutions to real life applications give each topic more meaning. Hopefully, you will be motivated and encouraged to learn even more.

### 4.4 Procedure:

The procedure section of each lab *outlines how to solve a problem.*

While it may not be apparent at first, technical textbooks introduce and discuss material in the same fashion as ***Steps 1 through 5 in Figure 2***.

A comparison of the seven procedure steps in ***Figure 2*** with any math and science textbook will reveal startling similarities.

Once you are aware and familiar with this method, you will find reading math and science textbooks much easier to understand. You will begin to know *what* to look for while you are reading.

Being aware of these seven steps will help you get more out of the sciences and get the most out of your learning experience.

An additional benefit of following the same steps outlined here is that you will find it much easier to write more efficiently about what you have learned. This will be especially useful at the high school and college levels.

**Steps 1 through 5** in the procedure sub-sections are referred to as the "***<u>theoretical results</u>*"**:

The theoretical result, while being very factual, will not provide the student with the "*<u>hands-on</u>*" application of the new concept. It is for this reason that steps 6 and 7 have been included in these labs.

**Steps 6 and 7** are referred to as the "***<u>empirical results.</u>***" Here, the new concept is applied and tested. The empirical results provide the "reality check" for the new concepts.

The new concepts are reinforced with a real application of the new material. Now, you can see for yourself that the concepts really work, and can be used for something tangible!

### 4.5 Results:

In each lab, you will have a chance to compare and evaluate the theoretical results with the empirical results. You will undoubtedly find there are differences in the two.

For example, a result using a calculator may be too precise if a ruler with 1/8-inch increments is used in the lab.

Rounding numbers in the theoretical evaluation may yield results too precise for the measurement in the lab. This may cause the results from the lab to differ from the theoretical results.

Differences in these two quantities will arise. However, they should be close enough to each other to give the student confidence that the theoretical work is a good estimate of the empirical results.

*When the two are **markedly** different, either the theoretical steps are flawed, or the lab was not conducted properly.*

*The goal is to give the student confidence in the theoretical method by reinforcing it with the empirical results.*

### 4.6 Conclusions:

*Did the math work?* Once the results are in, and comparisons can be made between the theoretical and empirical results, conclusions can be drawn.

A conclusion can be drawn from the entire process. Deviations, errors made in measurements, precision of the measurement instruments, etc., can be discussed.

***What you are looking for here is confidence that "the math works".***

In the conclusions portion of the lab, you may discover additional applications for what you learned. Share your thoughts with others!

For example, the results from the LAB 4 could be applied to the design of an automobile, or perhaps even a bicycle.

*See if you can find other unique applications for what you learned.*

### 4.7 Reference Section

In each lab, there is a reference section. The reference section will have useful resources for enquiring minds. Additional information on the topics discussed, web sites, books, and other resources are listed. Note: Web sites may change or may not be permanent.

### 4.8 INDEX

Each lab contains its own index. This may help you locate something you want to read again. Using the Table of Contents, the List of Figures and the Index are really good ways to navigate throughout the reading material, especially if you want to refer back to something in the book at a later date.

## 5 The Vertical Number Bar

The number bar centered between the columns is a meant to help you and the teacher address a location on a page you may want to discuss. For example, you may have a question located in the following part of the book:

| | |
|---|---|
| Book: | How Rockets Fly Grades 4-6 |
| Chapter: | LAB2: Creating Tables and Graphs to Determine Nosecone Lengths |
| Section: | 4-5 |
| Page: | 15 |
| Column: | Right |
| Position: | 12-20 |

## 6 Questions and Comments

If you have questions or comments regarding any of the material provided, please email:

aes2jwilson@fuse.com

or

***aes2jwilson@gmail.com***

*The word "**formula**" is commonly in use in science and math textbooks. It is strongly advised that the word "formula" **not be used**.*

*The word formula is a word reserved for chemists, and in the case of chemistry, is perfectly appropriate since the mixing of ingredients is what they do.*

*Instead, try replacing the word "**formula**" with the phrase*

***"a mathematical expression that defines a picture".***

*This will encourage you to look for the picture that a mathematical expression describes in any math or science textbook you read.*

*Remember, without a labeled picture, mathematical expressions are difficult to understand, if not meaningless.*

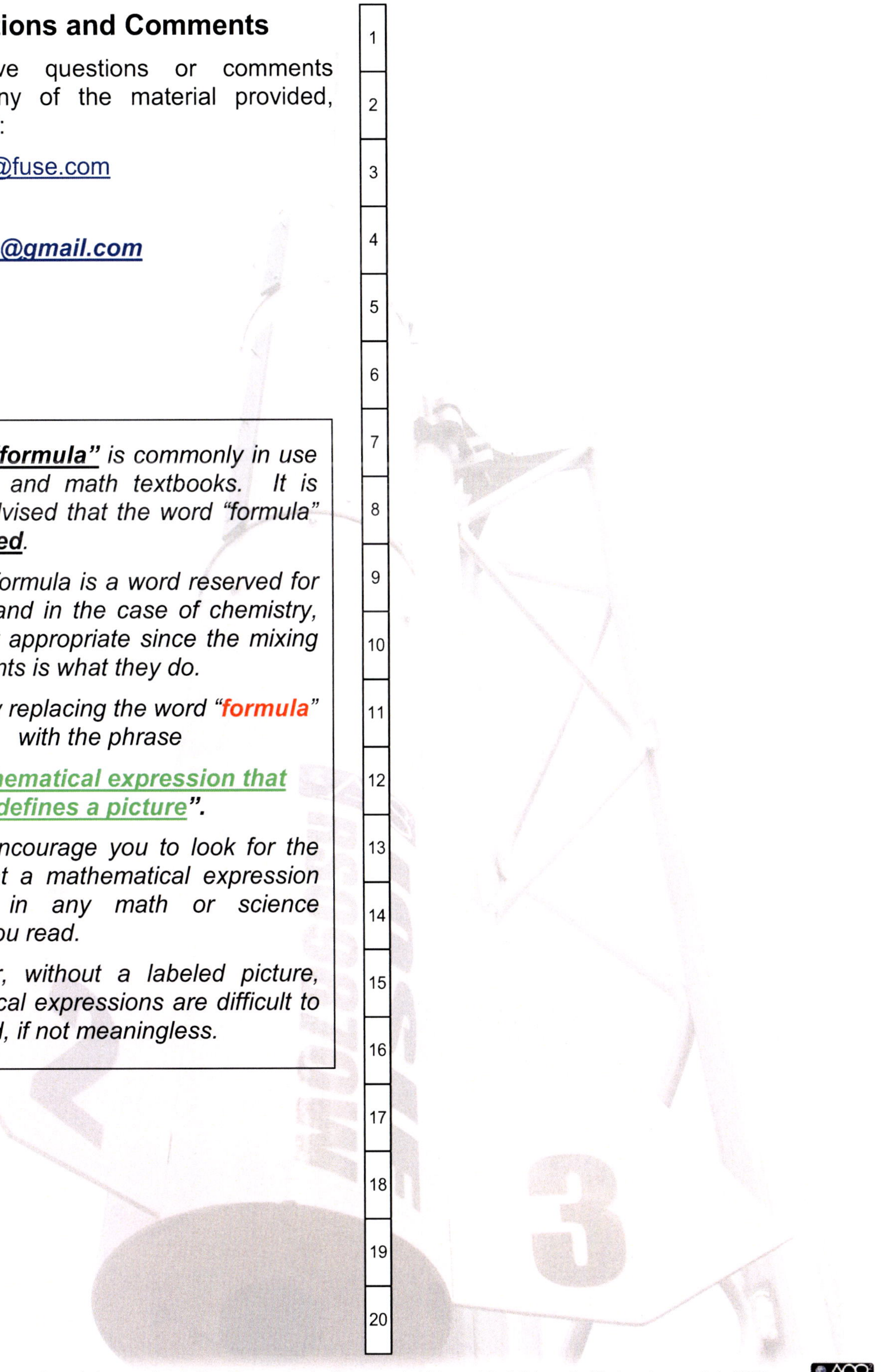

# LAB 1: CALCULATING NOSECONE LENGTHS

# LAB 1: CALCULATING NOSE CONE LENGTHS

4th - 6th Grade Learning Standards Applied to High Powered Rocket Flight and Descent

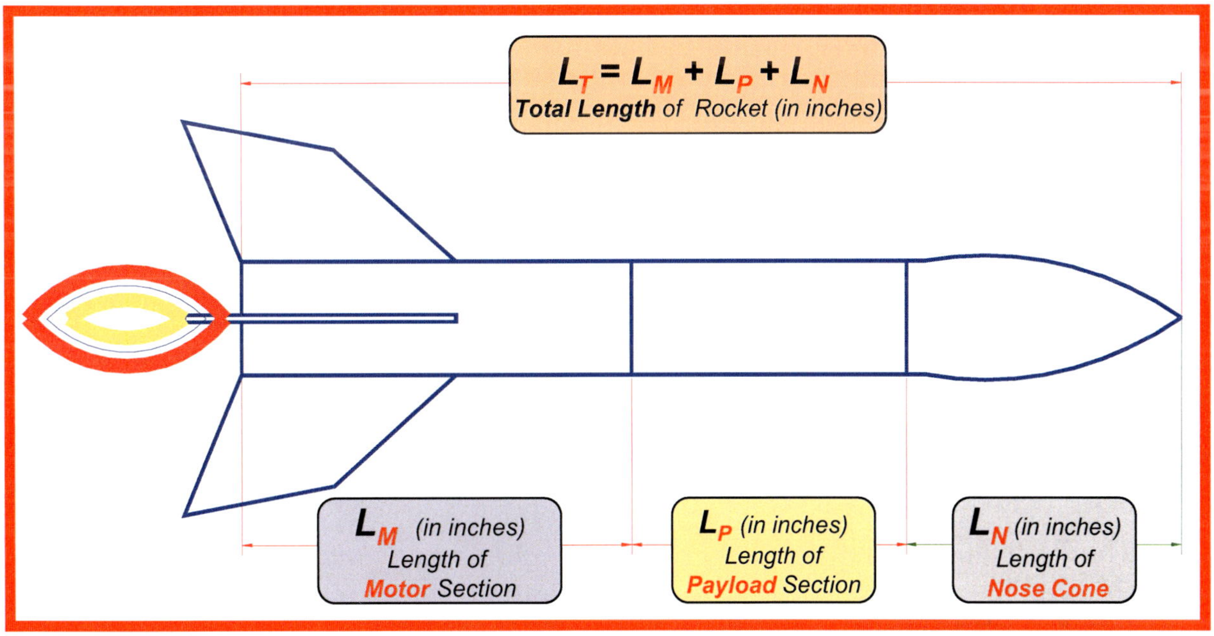

## Table of Contents

## List of Figures

## 1 Abstract (What are we going to learn?)

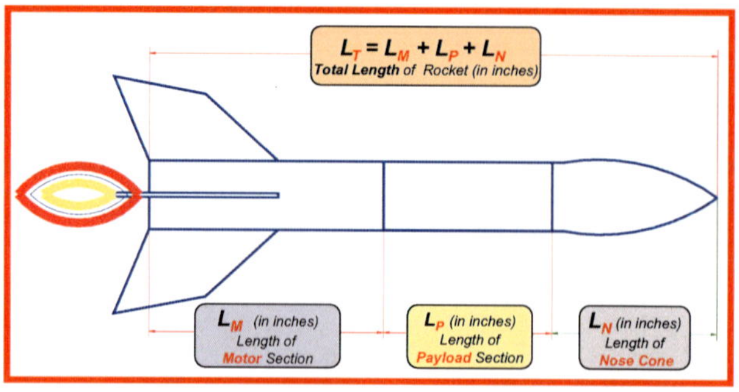


**Figure 1: Picture of our rocket**

In this lab, you will learn how long a nosecone will be based on the sizes of other parts of a rocket. We are going to accomplish this by creating "*a mathematical expression that defines a picture*." If we wanted to state the problem in words, it would go like this:

***"The total length of a rocket equals the length of the motor section plus the length of the payload section plus the length of the nosecone."***

Any problem written in words can be written with letters and symbols too. They both say the same thing, but with symbols, shorter amounts of space are used. Look at the mathematical expression below. It says the same thing as the sentence in green above.

$$L_T = L_M + L_P + L_N$$

Look at the picture in ***Figure 1***. Can you see the symbols? As you can see, it's a lot easier writing the mathematical equation than it is to write the words that describe the picture.

## 2 Glossary of Terms (New words and symbols we need to learn)

Here are the words and symbols we will use in this lab:

$L_T$: This stands for the **total length** of the rocket. $L_T$, the total length, has **units** of **inches**. Written with units it looks like this:

$$L_T\,(\text{in})$$

$L_M$: This stands for the **motor length** of the rocket. This is the part of the rocket that has the fins attached and the motor mounted inside. $L_M$, the motor length has **units** of **inches**. Written with units it looks like this:

$$L_M\,(\text{in})$$

$L_P$: This stands for the **payload length** of the rocket. This is the part of the rocket that has the parachute, maybe a camera, or a bug that wishes to go on a really scary, fast ride. $L_P$, the payload length, has **units** of **inches**. Written with units it looks like this:

$$L_P\,(\text{in})$$

$L_N$: This stands for the **nosecone length** of the rocket. This is the part of the rocket that moves air out of its path so it can go really, really fast. $L_N$, the nosecone length, has **units** of **inches**. Written with units it looks like this:

$$L_N\,(\text{in})$$

Whenever you start to read something in math or science, you have to pay attention to the symbols because they ALL have meaning. They say a lot in just a little space.

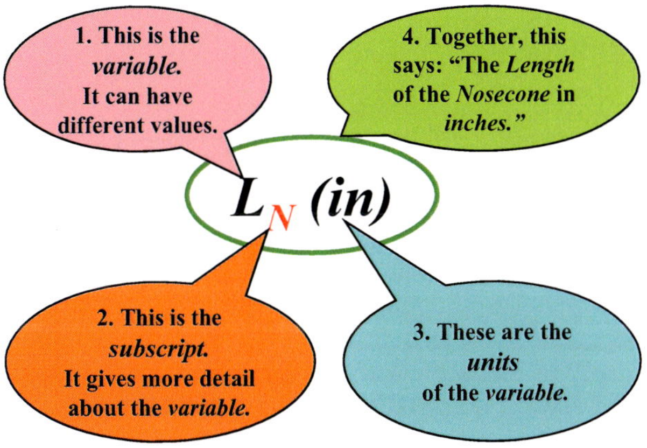


**Figure 2: Subscripts and UNITS**

Look at ***Figure 2***. Notice the small letters just below the symbol in **RED**. These are called **subscripts**. The subscripts are used to give more information to the symbol. Say the word "subscripts" a few times, so it doesn't seem like such an odd word to you.

We also have to pay attention to the **units**. In this lab, *everything is measured in **inches***. If you try to complete this lab and measure in centimeters or millimeters, we are going to end up with a really small rocket.

In LAB4 we will talk about something called a **superscript**. ***Figure 3*** is an example of a variable with a superscript, subscript and units.

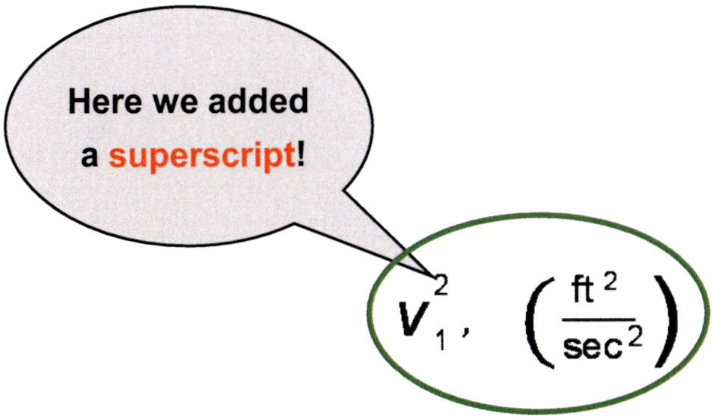


**Figure 3: Subscripts, superscripts and units**

## 3 Introduction and Purpose (Why are we doing this?)

The purpose of this lab is to introduce the student to using symbols to solve a problem. So, if we are going to learn something new, we might as well apply it to something we can make together. ***We are going find out how long we should make a nosecone for a rocket.***

In this problem, the motor and payload sections of the rocket have already been designed and built. We've been asked to provide a proper nosecone length **with the stipulation that the total rocket length when completed has to be 17 inches long**.

$$L_T = 17 \text{ inches}$$

We have to figure out how long to make the nosecone so that the rocket ends up being 17 inches long. When we find out how long the nosecone needs to be, we will draw the entire rocket on a piece of paper. Then we will measure our drawing to see if our overall length is 17 inches long. If it is, we can use this mathematical expression to build a real rocket!

## 4 Procedure: (How do we do it?)

1. Draw a picture of the problem
2. Label the picture
3. Write a mathematical expression that defines the picture
4. Solve for the unknowns
5. Calculate the results (by hand, by calculator, or by computer application software)
6. Create a lab to verify or deny the work in 1-5
7. Test the results of 1-5 in the lab

**Figure 4: The seven steps to solving any problem**

Okay. **Now remember, the way we solve this problem can be used to solve any problem**. It will seem very strange at first, but after you do a few of these, it will get much easier. We are going to solve this problem with **seven steps**. Read the steps in ***Figure 4.***

All of the Labs, including this Lab will follow the same steps in problem solving!.

### 4.1 Draw a picture of the problem

1. Draw a picture of the problem
2. Label the picture
3. Write a mathematical expression that defines the picture
4. Solve for the unknowns
5. Calculate the results (by hand, by calculator, or by computer application software)
6. Create a lab to verify or deny the work in 1-5
7. Test the results of 1-5 in the lab

Okay. Let's start out by drawing a picture of our rocket. We will draw one section at a time. **Make your own drawing on a separate piece of paper.** First, we will draw a picture of the **motor section** with its fins.

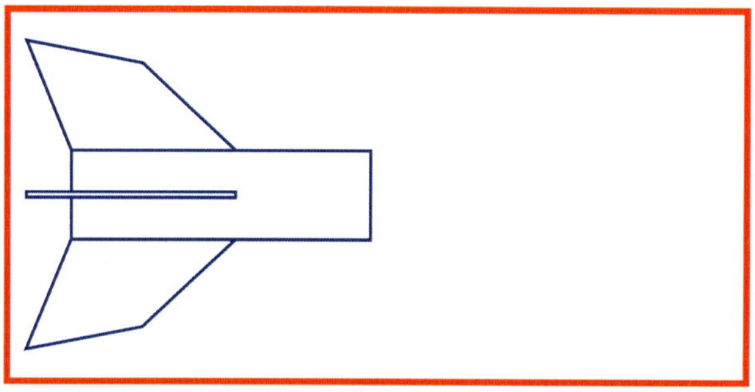

**Figure 5: Motor section with fins**

Next we will add the **payload section**. By the way, if you're thinking about putting your pet in the payload section, I should warn you that we are not allowed to do that. Let's just use the payload section **for our parachute for now.**

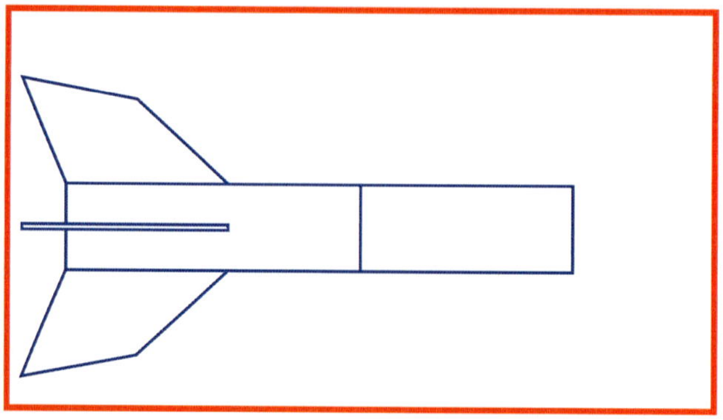

**Figure 6: Adding the payload section**

Now let's add the **nosecone**. We will discuss how important the shape of a nosecone is in a later Lab.

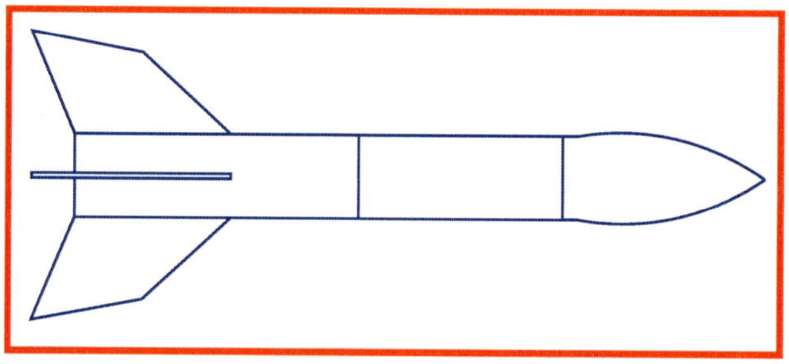

**Figure 7: Completed rocket picture**

Great! Now let's add some flames from the motor to make it look like its flying. We really don't need the **flames** for our problem, but it makes the picture look better.

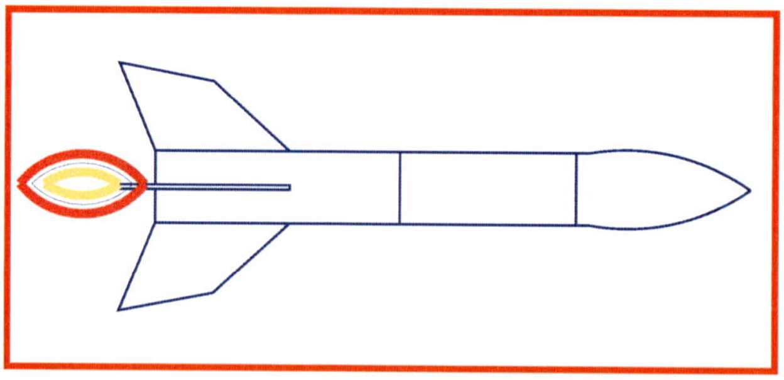

**Figure 8: Completed rocket with "cool" flames**

Well, there you have it. We've completed the first step to solving a problem. ***Drawing the picture is the most important step in solving a problem***. Now that we have drawn the picture, we can label it. Let's go ahead and do that.

## 4.2 Label the picture

1. Draw a picture of the problem
2. **Label the picture**
3. Write a mathematical expression that defines the picture
4. Solve for the unknowns
5. Calculate the results (by hand, by calculator, or by computer application software)
6. Create a lab to verify or deny the work in 1-5
7. Test the results of 1-5 in the lab

Now we are ready to label the picture. Soon you will be able to write your own math books, just the way you want them. First, let's label all the things we know. We know that the **total rocket length**, $L_T$, has to be **17 inches** long. Let's add that to the picture.

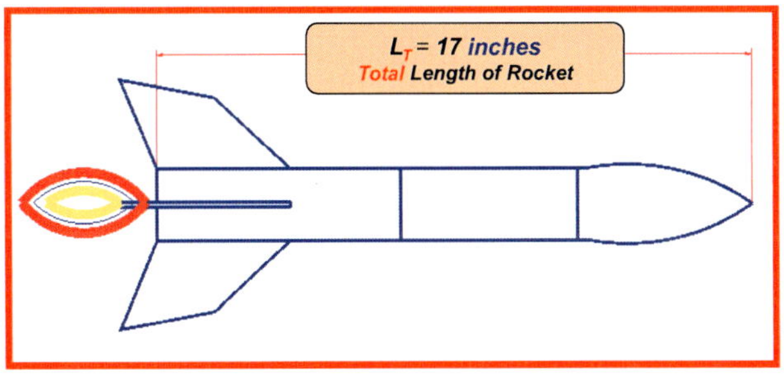


**Figure 9: Rocket with Total Length, $L_T$ label**

We have been told that the motor and payload sections have already been built. The **motor section length**, $L_M$, was built **7 inches** long. Let's add that to the picture.

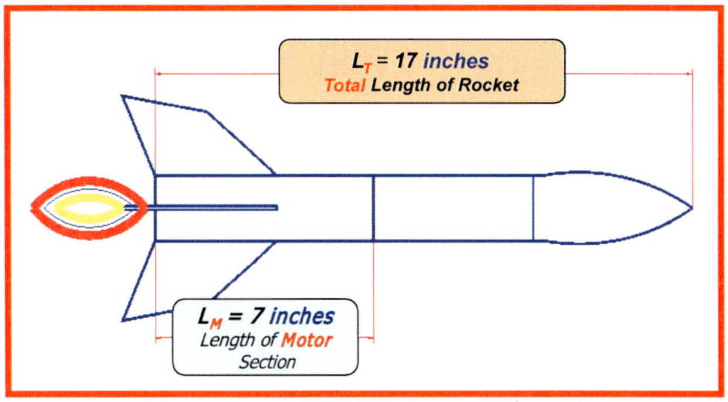


**Figure 10: Motor Section Length, $L_M$ label**

Remember, every time you read a math or science book, there probably will be a picture that represents the problem. If you don't see a picture, read the words and see if you can draw the picture yourself!

The **payload section length**, $L_P$, was built **5 inches** long. Let's add that to the picture.

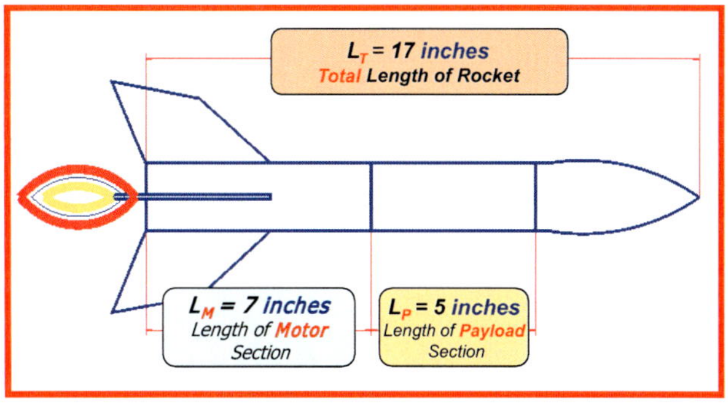


**Figure 11: Rocket with Payload Section Length, $L_P$, added**

Now the only thing we have to label is the **nosecone length**, $L_N$. We don't know how long it will be yet, but soon we will. Let's add the nosecone label to the picture. For now we'll just label our picture with a nosecone length "$L_N$ = **? inches**". We will solve for this value later and replace the question mark with our calculated value.

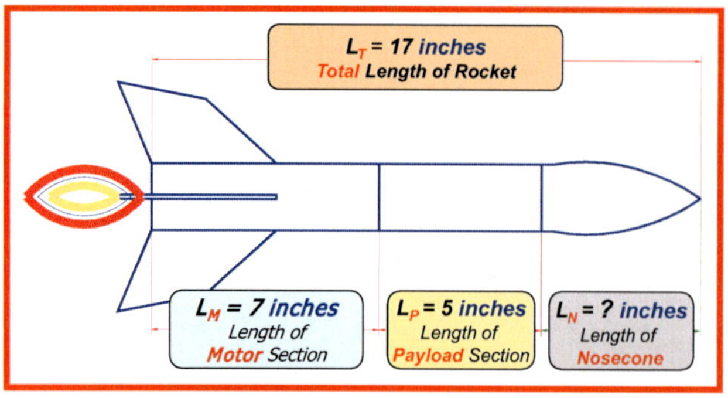


**Figure 12: Completed rocket picture!**

We are now ready to go on to the next step. Let's <u>write a mathematical expression that defines the picture</u>.

### 4.3 Write a mathematical expression that defines the picture

1. Draw a picture of the problem
2. Label the picture
3. **Write a mathematical expression that defines the picture**
4. Solve for the unknowns
5. Calculate the results (by hand, by calculator, or by computer application software)
6. Create a lab to verify or deny the work in 1-5
7. Test the results of 1-5 in the lab

Look at ***Figure 13***. Can you see why we can write the following sentence?

***"The total length of a rocket equals the length of the motor section plus the length of the payload section plus the length of the nosecone."***

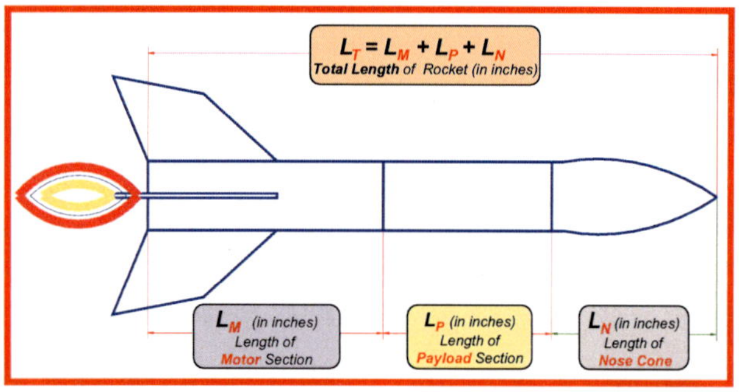


**Figure 13: Completed rocket picture!**

If we want to write the sentence in **symbols**, we can write it this way:

$$L_T = L_M + L_P + L_N$$

That's not so bad is it? This mathematical expression defines the picture we just drew and says that, if we add up the length of all the parts of the rocket, we will get its total length.

### 4.4 Solve for the unknowns

1. Draw a picture of the problem
2. Label the picture
3. Write a mathematical expression that defines the picture
4. **Solve for the unknowns**
5. Calculate the results (by hand, by calculator, or by computer application software)
6. Create a lab to verify or deny the work in 1-5
7. Test the results of 1-5 in the lab

The next step is **solving for the unknowns**. This may sound fairly tough, but as you will see, it's not nearly as bad as you might think. Our original mathematical expression that defined the picture was:

$$L_T = L_M + L_P + L_N$$

Since we don't know what the length of the **nosecone**, $L_N$, is supposed to be, we must ***<u>solve for it</u>***.

First, we can subtract the motor length, $L_M$, from both sides of our equal sign.

$$L_T - L_M = L_M - L_M + L_P + L_N$$

$$L_T - L_M = (L_M - L_M) + L_P + L_N$$

We can remove the $(L_M - L_M)$ because anything minus itself equals "0".

That leaves us with:

$$L_T - L_M = L_P + L_N$$

We can do the same thing with the length of the payload section, $L_P$.

$$L_T - L_M - L_P = (L_P - L_P) + L_N$$

Again, we remove the $(L_P - L_P)$ because anything minus itself equals "0".

That leaves us with the new symbol sentence:

$$L_T - L_M - L_P = L_N$$

In words:

*"The length of the nosecone, $L_N$, <u>equals</u> the total length of the **rocket**, $L_T$, <u>minus</u> the length of the motor section, $L_M$, <u>minus</u> the length of the **payload section**, $L_P$".*

Look at the picture in ***Figure 13*** and the new symbol sentence above. Compare the symbol sentence with the word sentence.

***<u>Take your time.</u>*** Do they say the same thing? There you have it. Now we can calculate the nosecone length. In the next lab, we will go through all of the steps that we used to arrive at our new mathematical expression.

We rearrange mathematical expressions by using *Algebra*. It's a funny name for the *rules* everyone uses to rearrange a mathematical expression.

Basically, the rules say that, if you change something on one side of the equal sign, and do the same thing to the other side of the equal sign, the mathematical expression says the same thing as it did before you made the change.

It may look different, but by using the rules, you can reverse your steps, and return to the original expression. If you can't wait until our next lab, ask your teacher to show you how it was done here with some other examples.

Now that we have completed <u>Steps 1 through 4</u>, we can find out how long our nosecone must be by just replacing the symbols with the lengths we already know. Use the lengths from the ***Figure 12***.

### 4.5 Calculate the results (by hand, by calculator, or by computer)

1. Draw a picture of the problem
2. Label the picture
3. Write a mathematical expression that defines the picture
4. Solve for the unknowns
5. **Calculate the results (by hand, by calculator, or by computer application software)**
6. Create a lab to verify or deny the work in 1-5
7. Test the results of 1-5 in the lab

At the beginning of our lab, we were given some rocket specifications. Let's have a look at those now and let's find out how long our nosecone needs to be.

**ROCKET SPECIFICATIONS:**

A rocket length that was ***17 inches*** long.

A motor section that was ***7 inches*** long.

A payload section that was ***5 inches*** long.

Our mathematical expression that defined our picture looked like this:

$$L_T - L_M - L_P = L_N$$

***Or***

$$L_N = L_T - L_M - L_P$$

Writing our mathematical expression that defines our picture with ***actual numbers*** looks like this.

$$L_N = 17\ (in) - 7\ (in) - 5\ (in)$$

***So:***

$$L_N = 5\ inches$$

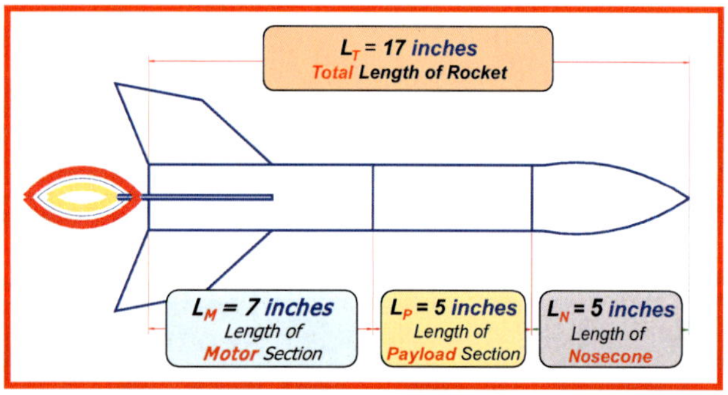


**Figure 14: Our rocket with the nosecone length added**

**What would the nosecone length have to be if the total length of the rocket had to be 15 inches long?**

________________________________________

________________________________________

In LAB4 we will discuss the diameters of a rocket and how it affects drag and flight performance. For now, we will use a rocket with a 2 inch diameter.

**What would the nosecone length have to be if the total length of the rocket was 19 inches long?**

________________________________________

________________________________________

In our next lab, we will make use of our mathematical expression to create a table. This table will be useful in designing rockets of varying sizes.

But for now, we want to set up a test to see if our math is correct. Let's go to the next step and see if we can prove that this length, 5 inches, is right, and that our work up to this point can be proven to be correct. We'll do this with a "lab".

**4.6 Create a lab to verify or deny the work in 1-5**

1. Draw a picture of the problem
2. Label the picture
3. Write a mathematical expression that defines the picture
4. Solve for the unknowns
5. Calculate the results (by hand, by calculator, or by computer application software)
6. **Create a lab to verify or deny the work in 1-5**
7. Test the results of 1-5 in the lab

Our picture represents the outline of a model rocket, but, as you can tell, the illustrations in this lab are not really **17 inches** long.

Let's draw all of the pieces of the rocket using a ruler, making sure all of the sections of the rocket are made to the measurements in our labeled drawing.

**We will draw each section of the rocket on a NEW piece of paper using the dimensions on the drawing.**

After we draw the pieces of the rocket, we will tape the pieces together. The total length of our rocket should measure 17 inches. Below is the procedure for our lab.

## LAB1 PROCEDURE:

1. Cut a piece of paper 7 inches long by 2 inches wide for the motor section.
2. Cut another piece of paper 5 inches long by 2 inches wide for the payload section.
3. Cut another piece of paper 5 inches long by 2 inches wide for the nosecone.
4. Once you have cut out the rectangle for the nosecone, fold it in half and cut a curved section out of this piece so it looks like a nosecone.
5. Make two fins to look like the one in our drawing.
6. Make a flame to come out the motor section.
7. Tape all the pieces together so it looks like our drawing.

### 4.7 Test the results of 1-5 in the lab.

1. Draw a picture of the problem
2. Label the picture
3. Write a mathematical expression that defines the picture
4. Solve for the unknowns
5. Calculate the results (by hand, by calculator, or by computer application software)
6. Create a lab to verify or deny the work in 1-5
7. **Test the results of 1-5 in the lab**

Once you have taped all the pieces together, carefully measure the *TOTAL LENGTH* of your rocket. Have someone else measure your rocket as well.

**How long was your rocket when you measured it? Measure your rocket to the nearest 1/16 of an inch.**

________________________________

________________________________

**Did you measure someone else's rocket? If so, how long was it? Measure their rocket to the nearest 1/16 of an inch.**

________________________________

________________________________

## 5 Results

You probably found out that the mathematical expression we used to solve for the length of the nosecone was right.

The overall length of the rocket should have been very close to **17 inches**. If you didn't get **17 inches** for the total length, your measured value was probably very close to **17 inches**.

The difference in what we predicted and what we actually measured is called **"error"**. It comes up often in real life applications. Several steps in our lab may have been the cause of an **error**.

**Some causes for the errors might be**:

- Your scissors may have been dull and you couldn't follow the outline of your drawing perfectly.
- Or, maybe when you taped the pieces together they shifted somewhat.
- Or, when you drew each of the sections of the rocket, they weren't drawn perfectly because your pencil wasn't sharp enough.

There may be other reasons for the errors, but the one thing to remember here is that even when using the most precise tools, there will always be *some* **error**.

This would be a good time to ask your teacher about the differences between **accuracy** and **precision**.

There's an example of the differences between **accuracy** and **precision** in ***Figure 15***.

## 6 Conclusions

The first few times you solve problems like this you may have the urge to take shortcuts, and, while that may work some of the time, the way you just solved this problem will **ALWAYS** work.

**The SEVEN STEPS we use to solve a problem are:**

1. Draw a picture of the problem
2. Label the picture
3. Write a mathematical expression that defines the picture
4. Solve for the unknown
5. Calculate answers (**theoretical answers**)

**Then we prove our calculated answers by:**

6. Creating a lab to test our theoretical answers
7. Performing the lab and testing the results

Once we have completed the lab, we can compare our lab results (**empirical answers**) with our calculated results (**theoretical answers**).

**REMEMBER:**

**Theoretical answers (Results): Predicted**

**Empirical answers (Results): Observed**

We are going to use the same mathematical expression we created here in the next lab. We will use it to design various sizes of rockets. The rockets will have the same basic parts. They will have a nosecone, a payload section, and a motor section.

What's great about writing a mathematical expression is that once a mathematical expression is created, and proven to be correct, it can used over and over again!

*Go back to the beginning of this lab and read the abstract. Does it make more sense now?*

**Can you find other uses for what you learned in this lab?**

________________________________________

________________________________________

________________________________________

________________________________________

## 7 References and additional reading


**7.1 The Handbook of Model Rocketry by G. Harry Stine**

**7.2 The Parachute Manual by Dan Poynter**

**7.3 National Association of Rocketry,** www.nar.org

**7.4 Email:** nar-hq@nar.org

**7.5** http://www.nasa.gov

**7.6** http://www.Spacex.com

# 8 INDEX



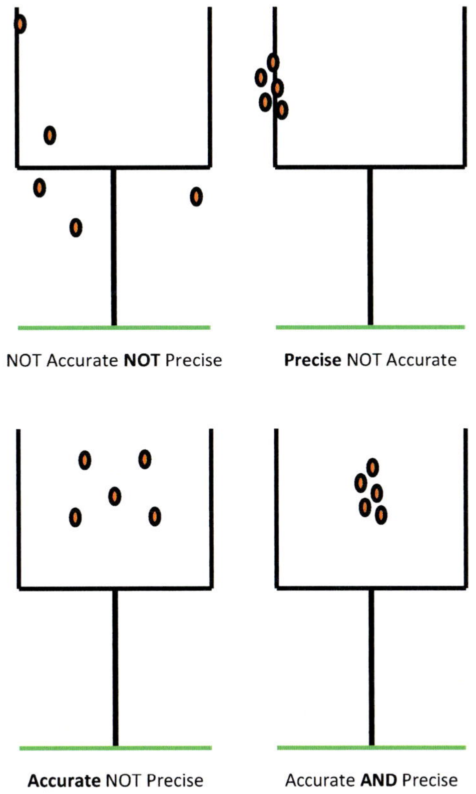


**Figure 15: The difference between accuracy and precision**

# LAB 2: CREATING TABLES & GRAPHS

# LAB 2: CREATING TABLES & GRAPHS TO DETERMINE NOSECONE LENGTHS

4th - 6th Grade Learning Standards Applied to High Powered Rocket Flight and Descent

| Rocket # | Rocket Total Length $L_T$ inches Given to us | Length of the Payload Section $L_P$ inches Given to us | Length of the Motor Section $L_M$ inches Given to us | Length of the Nose Section $L_N$ inches We calculate |
|---|---|---|---|---|
| | | | | $L_N = L_T - L_P - L_M$ |
| 1 | 15 | 7 | 5 | |
| 2 | 17 | 7 | 5 | |
| 3 | 19 | 7 | 5 | |
| 4 | 21 | 7 | 5 | |
| 5 | 23 | 7 | 5 | |
| 6 | 25 | 7 | 5 | |
| 7 | 27 | 7 | 5 | |
| 8 | 29 | 7 | 5 | |
| 9 | 31 | 7 | 5 | |
| 10 | 33 | 7 | 5 | |

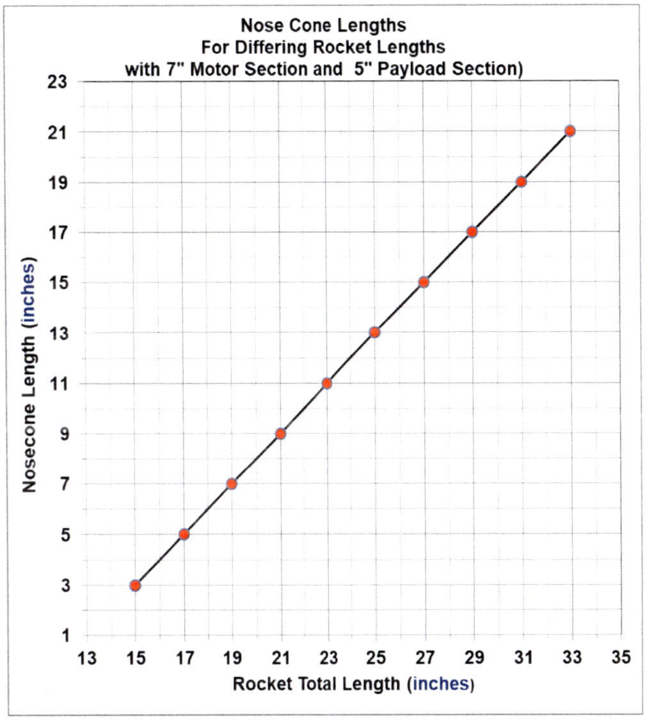


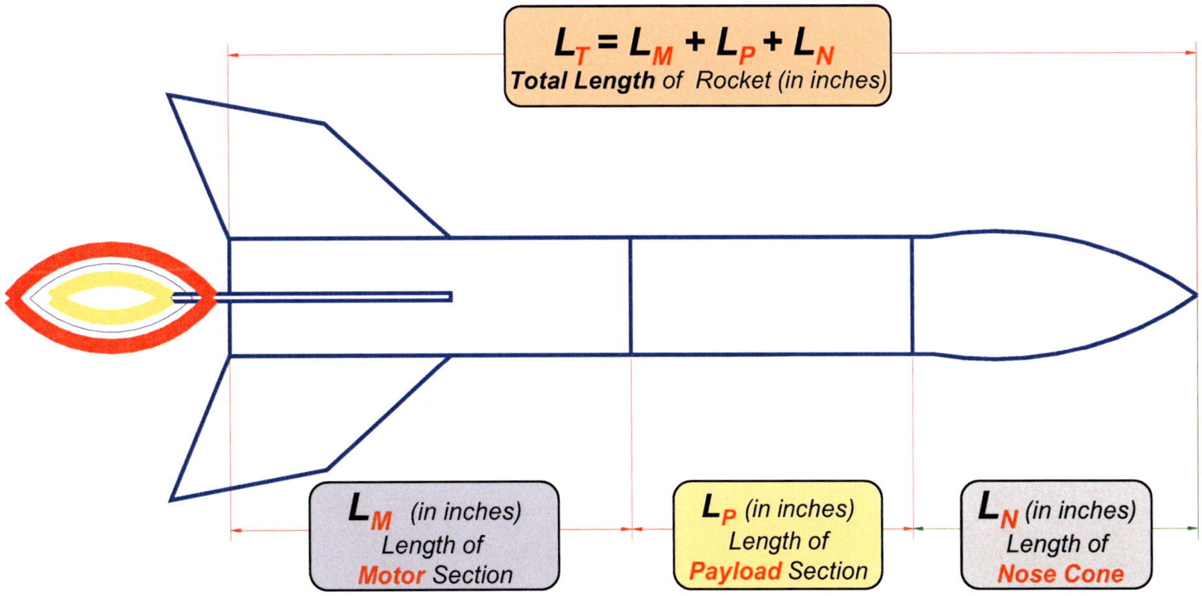

## Table of Contents

## List of Figures

## 1 Abstract (What are we going to learn?)

Tables and graphs are created from mathematical expressions.

In this lab, we are going to use our mathematical expression to make a **table**. From the table, we are going to make a **graph**. Since we already have our mathematical expression, we are going to put the graph to good use to design rockets of different lengths.

In LAB 1, we learned that we could create a sentence from a picture and turn it into a mathematical expression that described the picture. Our sentence went like this:

"The total length of a rocket equals the length of the motor section plus the length of the payload section plus the length of the nosecone."

The sentence was then written in symbols (a symbol sentence) to form the mathematical expression below.

$$L_T = L_M + L_P + L_N$$

Later, we solved for the **unknown,** the length for a **nosecone**. Then we calculated the nosecone length based on the total rocket length, the payload length and the motor length.

$$L_N = L_T - L_P - L_M$$

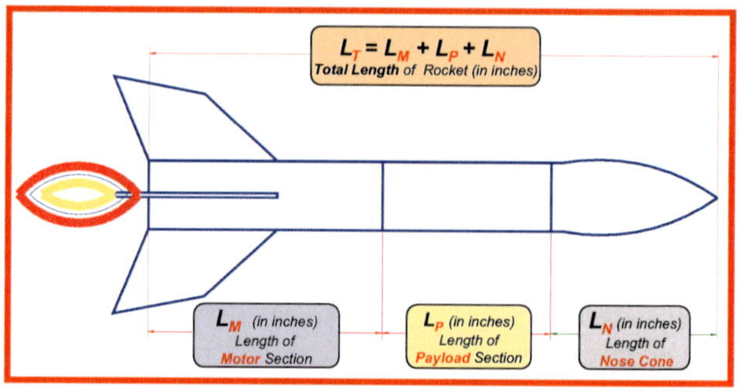


**Figure 1: Picture of our rocket**

## 2 Glossary of Terms (New words and symbols)

**Abscissa axis:** This is a term used to describe the axis on a graph that is normally horizontal. The numbers on this axis determine the numbers on the *ordinate* axis.

**Ordinate axis:** This is a term used to describe the axis on a graph that is normally vertical. The numbers on the *abscissa* axis determine the numbers on this axis. **In other words, the ordinate numbers are a function of the abscissa values!**

### Review from LAB1:

$L_T$: This stands for the **total length** of the rocket. $L_T$, the total length, has **units** of **inches**. Written with units it looks like this:

$$L_T\text{(in)}$$

$L_M$: This stands for the **motor length** of the rocket. This is the part of the rocket that has the fins attached and the motor mounted inside. $L_M$**,** the motor length has **units** of **inches**. Written with units it looks like this:

$$L_M\text{(in)}$$

***$L_P$***: This stands for the **payload length** of the rocket. This is the part of the rocket that has the parachute, maybe a camera, or a bug that wishes to go on a really scary, fast ride. ***$L_P$***, the payload length, has **units** of **inches**. Written with units it looks like this:

$$L_P\,(in)$$

***$L_N$***: This stands for the **nosecone length** of the rocket. This is the part of the rocket that moves air out of its path so it can go really, really fast. $L_N$, the nosecone length, has **units** of **inches**. Written with units it looks like this:

$$L_N\,(in)$$

Yak: a long-haired ox that lives in central Asia.

## 3 Introduction and Purpose (Why are we doing this?)

The purpose of this lab is to introduce methods used to create both a number table and a graph. These are created using mathematical expressions that defined a picture.

Tables of information are all over the place, and so are graphs. Tables and graphs are created using mathematical expressions. We all gather and learn information and concepts in different ways. Some people prefer to *read* the information about a new topic; others like to see the *mathematical expression* that defines a picture, while others prefer to look at *tables and graphs.*

Tables and graphs are great tools that illustrate many solutions of a problem in a single picture.

We go to a lot of effort creating a mathematical expression, but once we've completed, we can use it over and over.

By *varying only* ***one*** value, (the variable), in a mathematical expression, like the total rocket length, while leaving all but one of the other values alike (*constant*), we can solve for one of the other parts. We could solve the equation for the motor section, payload section or nosecone. We can then create a table and a graph that will represent our different solutions.

Remember the saying "***a picture is worth a thousand words***." That's why we always start with a picture (*the picture of the rocket for example*), and end with a picture of a table and a graph that represents our solutions of our mathematical expression.

When reading tables and graphs, the one thing to remember is that the information on the table and the graph came from a mathematical expression that defined a picture.

In this LAB, we will be using odd numbered values for the total length of a rocket. The lengths of the payload section and motor section will be the same for all of our rocket designs.

- We will vary, (change) the total length of the rocket.
- We will keep the motor section and payload sections constant, (the same) lengths.
- We will then determine, (calculate) the length for the nosecone for each rocket.
- Once we calculate the nosecone lengths based on the different rocket lengths and fixed values for the payload and motor sections, we will tabulate our results and create our own table that illustrates our results.
- Finally, we will create a graph based on the information in our table.

In this lab, we have been asked by the rocket designers to calculate the nosecone lengths for 10 different rockets.

It's our job to figure out how long to make each nosecone for each rocket so that the rocket has the total length the designers are asking for.

Since we already have the mathematical expression that defines our rocket, this will be easy and fun. The following list shows how long each rocket will be.

| Rocket # | The total length of the rocket, $L_T$, in inches Given to us |
|---|---|
| 1 | 15 |
| 2 | 17 |
| 3 | 19 |
| 4 | 21 |
| 5 | 23 |
| 6 | 25 |
| 7 | 27 |
| 8 | 29 |
| 9 | 31 |
| 10 | 33 |

**Figure 2: Ten *total lengths of rockets* given to us by the rocket designers**

**Remember, the motor and payload sections have already been built.**

**It is our job to figure out the nosecone lengths for each rocket**

## 4 Procedure: (How do we do it?)

1. **Draw a picture of the problem**
2. **Label the picture**
3. **Write a mathematical expression that defines the picture**
4. **Solve for the unknowns**
5. **Calculate the results (by hand, by calculator, or by computer application software)**
6. **Create a lab to verify or deny the work in 1-5**
7. **Test the results of 1-5 in the lab**

**Figure 3: The seven steps to solving any problem**

Okay. Now remember, we are going to solve this problem just as we did in LAB 1. We are going to solve this problem with our seven steps. Read the steps in ***Figure 3***.

As you will see, most of the work required to solve this problem was done in LAB 1. If you have LAB 1 handy, you may want to refer to it from time to time as you read this lab.

### 4.1 Draw a picture of the problem

1. **Draw a picture of the problem**
2. Label the picture
3. Write a mathematical expression that defines the picture
4. Solve for the unknowns
5. Calculate the results (by hand, by calculator, or by computer application software)
6. Create a lab to verify or deny the work in 1-5
7. Test the results of 1-5 in the lab

In the first lab, we drew a picture of the rocket, so that part is complete.

**The picture is the most important thing to do** so if you don't remember; you can always go back and review LAB1.

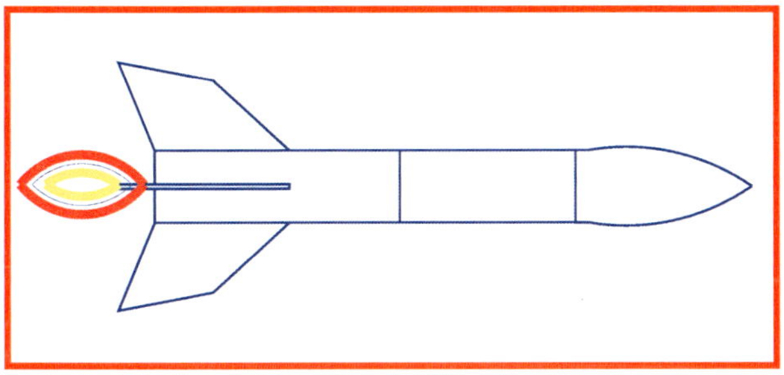

Figure 4: Completed rocket with cool flames

## 4.2 Label the picture

1. Draw a picture of the problem
2. **Label the picture**
3. Write a mathematical expression that defines the picture
4. Solve for the unknowns
5. Calculate the results (by hand, by calculator, or by computer application software)
6. Create a lab to verify or deny the work in 1-5
7. Test the results of 1-5 in the lab

Since we have already drawn the picture and labeled it, step 2 is also complete. ***Figure 5*** is the picture we labeled in LAB 1.

***Remember**, **every time you read a math or science book, there will probably be a picture that represents the problem***.

Look for the picture that the topic is describing. Then carefully read the description as you study the picture. If you don't see a picture, read the words. See if you can draw and label a picture yourself!

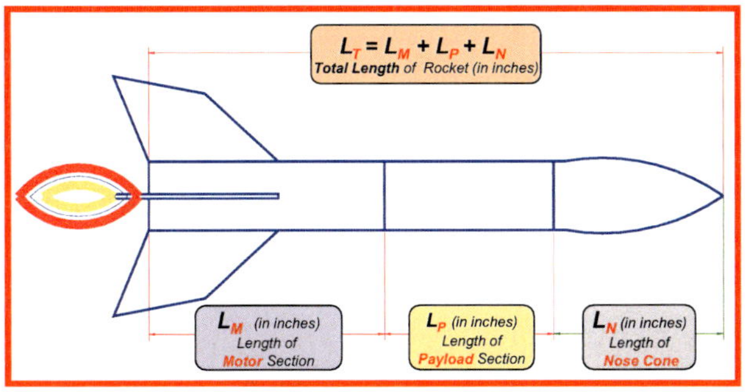


Figure 5: Completed rocket picture!

We are now ready to go on to the next step. Let's write a mathematical expression that defines the picture.

## 4.3 Write a mathematical expression that defines the picture

1. Draw a picture of the problem
2. Label the picture
3. **Write a mathematical expression that defines the picture**
4. Solve for the unknowns
5. Calculate the results (by hand, by calculator, or by computer application software)
6. Create a lab to verify or deny the work in 1-5
7. Test the results of 1-5 in the lab

Look at the labeled picture of the rocket in **Figure 5**. Can you see why we can write the following sentence?

*"The total length of a rocket equals the length of the motor section plus the length of the payload section plus the length of the nosecone."*

Take your time and see if you can tell how this ***symbol sentence*** below can be written from the word sentence above.

Do you remember how we did this in LAB1?

$$L_T = L_M + L_P + L_N$$

So, if you look closely at the picture that has been labeled, you can create both the word and symbol sentence.

As was mentioned in the LAB1, the symbol sentence is a shorthand version of the word sentence. The symbol sentence can be written much faster than the word sentence and it takes up less space!

## 4.4 Solve for the unknowns

1. Draw a picture of the problem
2. Label the picture
3. Write a mathematical expression that defines the picture
4. **Solve for the unknowns**
5. Calculate the results (by hand, by calculator, or by computer application software)
6. Create a lab to verify or deny the work in 1-5
7. Test the results of 1-5 in the lab

The next step is solving for the unknowns. Sounds pretty tough but, as you will see, it's not nearly as bad as you might think. Our original mathematical expression that defined the picture was:

$$L_T = L_M + L_P + L_N$$

Since we don't know what the length of the nosecone, $L_N$, is supposed to be, we solve for it.

### In words:

"The length of the Nosecone, $L_N$, equals the Total length $L_T$, of the rocket minus the length of the Motor, $L_M$ section minus the length of the Payload section, $L_P$."

We can write the word sentence with **symbols.** It looks like this:

$$L_N = L_T - L_P - L_M$$

How did we get from the ***symbol sentence*** box in green to the ***symbol sentence*** box in red?

Let's review the steps from LAB1:

*Starting with our original expression that defined our picture we had:*

$$L_T = L_M + L_P + L_N$$

*If we subtract $L_M$ from both sides of the equation, it looks like this:*

$$L_T - L_M = L_M - L_M + L_P + L_N$$

*If we subtract $L_P$ from both sides of the equation, it looks like this:*

$$L_T - L_M - L_P = L_M - L_M + L_P - L_P + L_N$$

*If we group the $L_M - L_M$ and $L_P - L_P$ symbols on the right hand side of the equal sign, it looks like this:*

$$L_T - L_M - L_P = (L_M - L_M) + (L_P - L_P) + L_N$$

*Since $(L_M - L_M) = 0$ and $(L_P - L_P) = 0$:*

*we can replace both of these groups with zeros in our expression. When we do that, it looks like this:*

$$L_T - L_M - L_P = (0) + (0) + L_N$$

*If we remove the zeros, our expression ends up like this:*

$$L_T - L_M - L_P = L_N$$

There you have it! ***The Length of the Nosecone can now be calculated.*** Ask your teacher how this method can be used in other ways.

## 4.5 Calculate the results (by hand, by calculator, or by computer application software)

1. Draw a picture of the problem
2. Label the picture
3. Write a mathematical expression that defines the picture
4. Solve for the unknowns
5. **Calculate the results (by hand, by calculator, or by computer application software)**
6. Create a lab to verify or deny the work in 1-5
7. Test the results of 1-5 in the lab

Now that we have done steps 1 through 4, we can find out how long our nosecones will have to be. Using our mathematical expression that defines our picture, we can calculate all of our nosecone lengths and create a table.

Let's make a ***table*** of our lengths and then make a ***graph*** from the table.

If we start with the first value from our list of rocket sizes, payload and motor section sizes, and use the mathematical expression for the nosecone, we can start to make a list of our answers. Below, **the first two have been solved and placed in the $L_N$ column** of our table.

$$L_N = L_T - L_P - L_M$$

**If the total length is 15 inches:**

$L_N$ = 15 inches – 7 inches – 5 inches

OR (much easier to write):

$L_N$ = (15 – 7 – 5) inches

$L_N$ = 3 inches

**If the total length is 17 inches:**

$L_N$ = (17 – 7 – 5) inches

$L_N$ = 5 inches

Continue down the $L_T$ column in ***Figure 6*** until the $L_N$ column, (rows 1-10), is complete.

Once we solve for all ten nosecone lengths in our list we will end up with enough information to create our own graph!

| Rocket # | Total Length of the Rocket $L_T$ inches Given to us | Length of the Payload Section $L_P$ inches Given to us | Length of the Motor Section $L_M$ inches Given to us | Length of the Nose Section $L_N$ inches We calculate |
|---|---|---|---|---|
| | | | | $L_N = L_T - L_P - L_M$ |
| 1 | 15 | 7 | 5 | 3 |
| 2 | 17 | 7 | 5 | 5 |
| 3 | 19 | 7 | 5 | |
| 4 | 21 | 7 | 5 | |
| 5 | 23 | 7 | 5 | |
| 6 | 25 | 7 | 5 | |
| 7 | 27 | 7 | 5 | |
| 8 | 29 | 7 | 5 | |
| 9 | 31 | 7 | 5 | |
| 10 | 33 | 7 | 5 | |

**Figure 6: Table of nosecone lengths**

Now that we have made a table of our calculations, we can make a graph that represents the table.

Let's go to the next step and make a graph. There are actually **10 steps** to making a graph.

It may seem like a lot of steps at first, but as you create more graphs, you'll see that the steps become automatic and go really fast. Review the steps for creating a graph in ***Figures 7 & 8***.

1. Draw the abscissa.
2. Put the values on the abscissa (from the table).
3. Label the abscissa including units.
4. Draw the ordinate.
5. Put the values on the ordinate from the table.
6. Label the ordinate including units.
7. Put a *Title* or Label on the graph.
8. Add vertical and horizontal lines to make the graph easier to read.
9. Put a point (dot) at each location from the table for every $L_T$ and every matching (same row from the table) $L_N$ from our table.
10. Draw a line connecting the points to make one continuous line.

**Figure 7: Ten steps to drawing a graph**

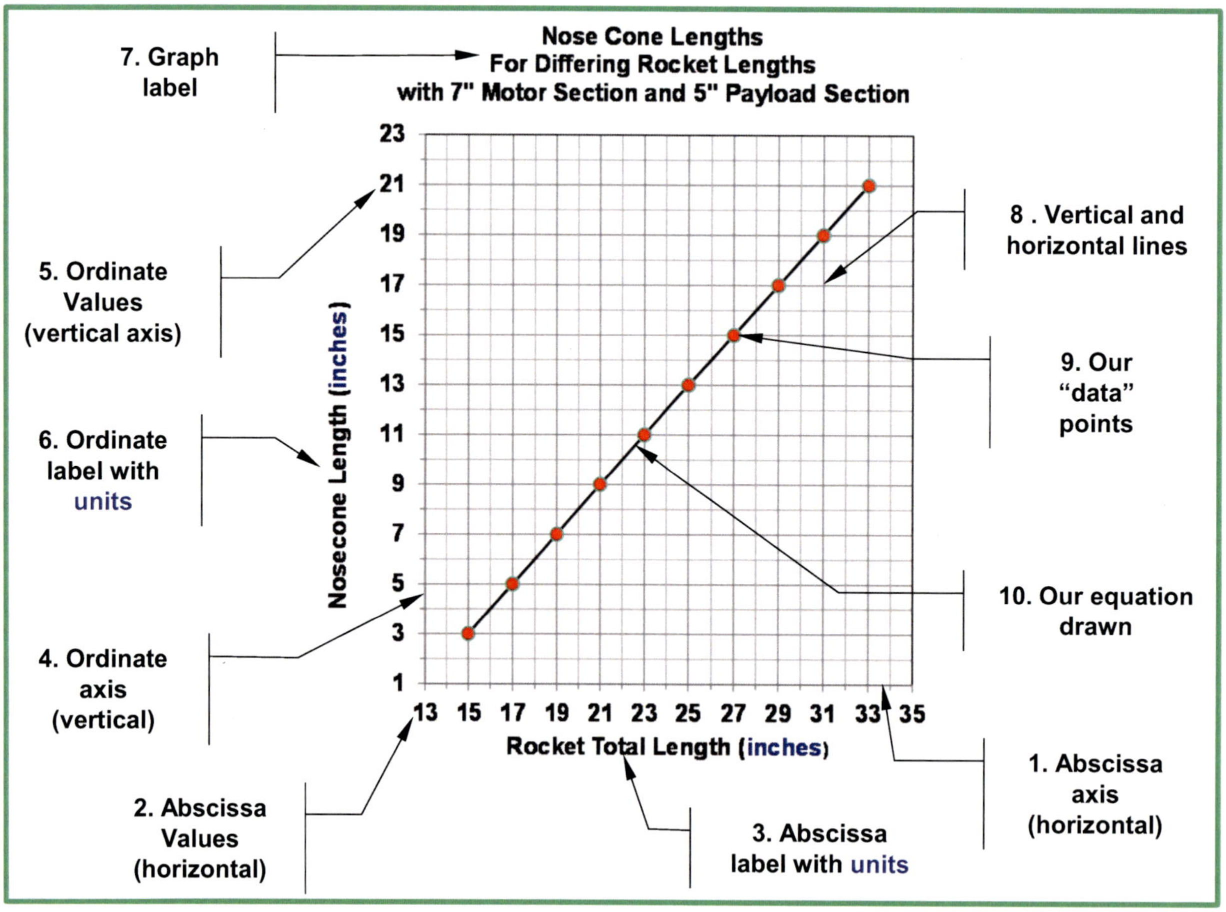


**Figure 8: Graph of our equation and the 10 steps needed to draw a graph**

***Figure 7*** shows a list of the **10 STEPS for drawing a graph**. If you skip any of these steps, the graph probably won't make any sense. So let's go through the steps one by one together.

Graph paper is provided throughout the book just for this purpose!

One thing that is very handy to have when making graphs is a special paper called ***graph paper***. It makes drawing graphs easier because it has equally spaced horizontal and vertical lines that form a ***grid***. This type of paper makes drawing straight lines on the graph a lot easier. *You don't have to use graph paper for our work here, but if you like, go ahead and use the graph paper provided throughout this book.*

As we continue, we'll go through each step and draw our own graph, step by step. Refer back to ***Figure 8*** as you go each of the 10 steps.

As you look at the graph in ***Figure 8***, you'll see there are little notes that describe each of the 10 steps in creating a graph. These are the same ten steps shown in the table (***Figure 7)***.

Compare the **graph** with the 10 steps in ***Figure 7*** and see if all of the steps make sense.

You'll find reading graphs made by other people much easier to understand once you've finished drawing your own graph.

**Step 1**:

Make a horizontal line and divide it up into equally spaced little "tic" marks. Our table has 10 nosecone lengths so we should have at least 10 "**tic**" marks on our abscissa axis. Look at ***Figure 9***.

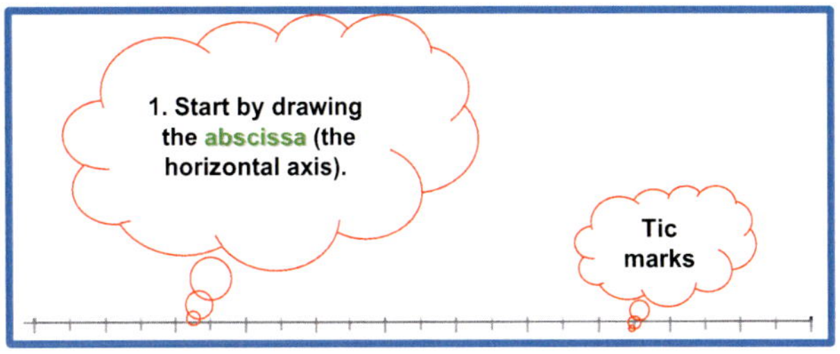


**Figure 9: (Step 1 of 10) Draw the abscissa axis**

**Step 2**:

Start with the first column of our table where our **Rocket Total Lengths** are listed. Start from the top of the column and work your way down to the bottom.

Put the Total Length, $L_T$, numbers from our table on the **abscissa** axis.

The top number from the table will be on the left side of the axis and the bottom number from the table will be on the right side of the axis

Sometimes it's convenient to add an extra number at each end of the abscissa. If you look at ***Figure 11***, you'll see how that's done.

| Rocket # | Rocket Total Length $L_T$ inches Given to us | Length of the Payload Section $L_P$ inches Given to us | Length of the Motor Section $L_M$ inches Given to us | Length of the Nose Section $L_N$ inches We calculate |
|---|---|---|---|---|
| | | | | $L_N = L_T - L_P - L_M$ |
| 1 | 15 | 7 | 5 | 3 |
| 2 | 17 | 7 | 5 | 5 |
| 3 | 19 | 7 | 5 | 7 |
| 4 | 21 | 7 | 5 | 9 |
| 5 | 23 | 7 | 5 | 11 |
| 6 | 25 | 7 | 5 | 13 |
| 7 | 27 | 7 | 5 | 15 |
| 8 | 29 | 7 | 5 | 17 |
| 9 | 31 | 7 | 5 | 19 |
| 10 | 33 | 7 | 5 | 21 |

Abscissa Values

**Figure 10: (Step 2 of 10) $L_T$ values from our table that go on the abscissa axis**

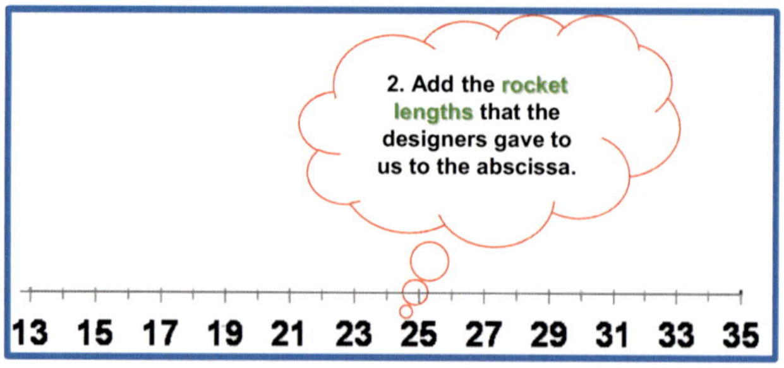


**Figure 11: (Step 2 of 10) Adding the numbers, $L_T$, from our table onto the abscissa**

**Step 3**:

Label the ***abscissa*** axis. This is where you tell the person reading the graph what the numbers on the abscissa represent. In this graph, they represent the overall **lengths of our rockets, in inches.** ***It is very important to include the units here***. Without units, no one reading the graph will be able to make sense out of it. *Without units, the reader might think the numbers represent the number of* ***yaks*** *in the Cincinnati Zoo!*

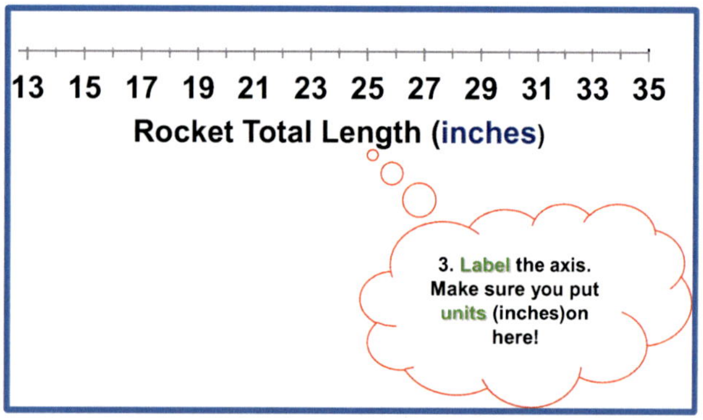


**Figure 12: (Step 3 of 10) Label the abscissa axis with the proper units**

**Step 4**:

Draw the ordinate axis just as you drew the abscissa axis. Put **equally** spaced tic marks on the ordinate axis like you did on the abscissa axis.

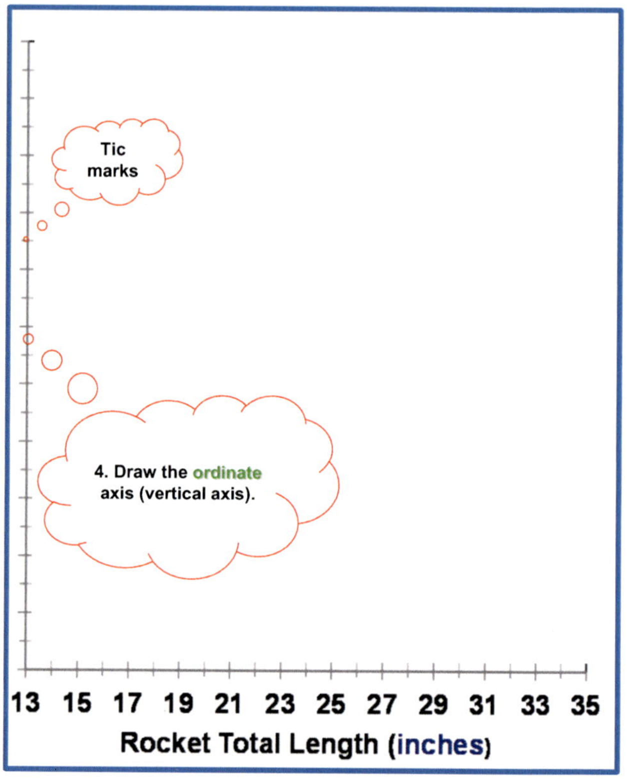


**Figure 13: (Step 4 of 10) Draw the ordinate (vertical axis)**

**Step 5**:

Okay, now let's put the numbers on the **ordinate** axis from our table. Start with the last column of our table where our **nosecone lengths** are listed.

Start from the top of the column and work your way down to the bottom. The top number from the table goes on the bottom of the axis and the bottom number from the table goes on the top of the axis.

Notice that there are extra numbers on our ordinate that weren't on the table. That's okay, and, as you will see later, we're going to use those numbers. What's important here is that you include the smallest and the largest number from our table on the ordinate axis and divide the space between these two numbers into equally spaced tic marks.

| Rocket # | Rocket Total Length $L_T$ inches Given to us | Length of the Payload Section $L_P$ inches Given to us | Length of the Motor Section $L_M$ inches Given to us | Length of the Nose Section $L_N$ inches We calculate |
|---|---|---|---|---|
| | | | | $L_N = L_T - L_P - L_M$ |
| 1 | 15 | 7 | 5 | 3 |
| 2 | 17 | 7 | 5 | 5 |
| 3 | 19 | 7 | 5 | 7 |
| 4 | 21 | 7 | 5 | 9 |
| 5 | 23 | 7 | 5 | 11 |
| 6 | 25 | 7 | 5 | 13 |
| 7 | 27 | 7 | 5 | 15 |
| 8 | 29 | 7 | 5 | 17 |
| 9 | 31 | 7 | 5 | 19 |
| 10 | 33 | 7 | 5 | 21 |

Ordinate Values

**Figure 14: (Step 5 of 10) Grab the ordinate values from our table for our graph**

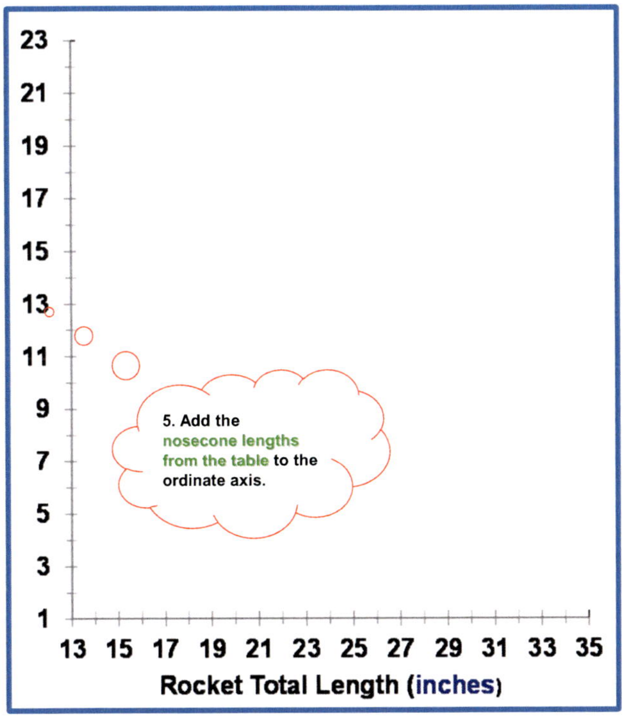


**Figure 15: (Step 5 of 10) Add the ordinate numbers on the graph (from the table)**

**Step 6:**

**Label** the **ordinate** axis. This is where you tell the person reading the graph what the numbers on the ordinate represent. In our case, they represent the **lengths of our nosecones**. It is very important to put the **units** here. Without units, the reader might think the numbers represent the pounds of hay a yak will eat for breakfast!

Ask your teacher how many pounds of hay a yak eats for breakfast. What is a yak anyway?

**We're almost done!**

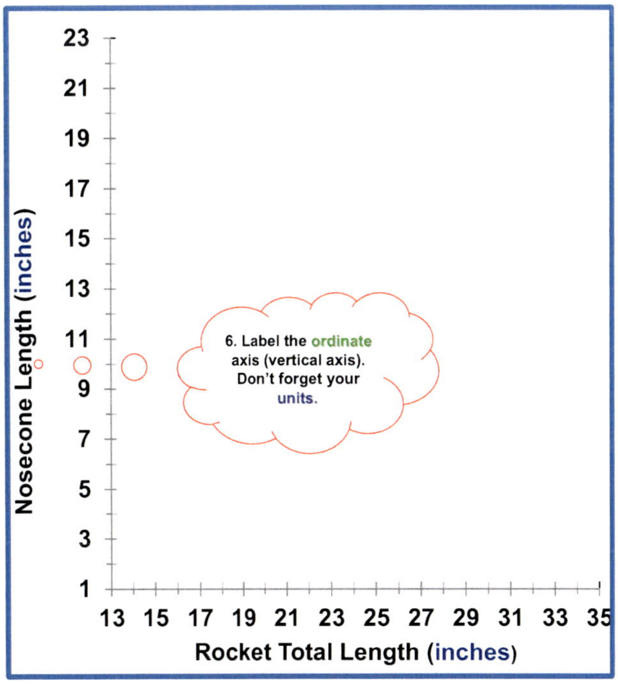


**Figure 16: (Step 6 of 10) Label the ordinate axis**

**Step 7**:

In step 7, we will put a **Title** on our graph. Any special conditions that we need to put on the graph are put here. If the values for the motor section or the payload section change, we have to create a new table and graph based on the changed values. For example, what would our nosecone length be if the payload section were 6 inches instead of 5 inches for each of the rockets? Would our nosecone lengths be different?

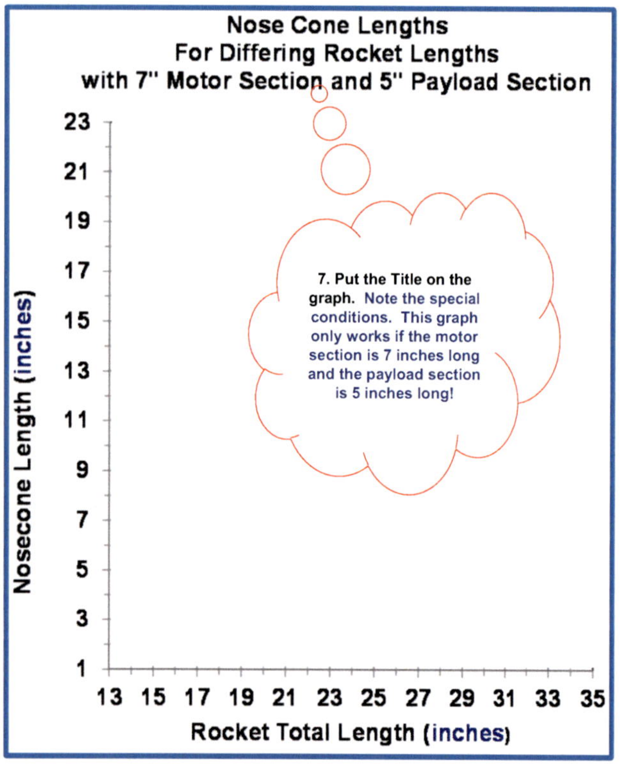


**Figure 17: (Step 7 of 10) Put a title on the graph**

Okay, now that we've labeled our graph, let's put some vertical and horizontal lines on our graph so we can put our data on the graph.

**Step 8:**

**If you are using graph paper, you can skip Step 8.**

In order for us to put our data points on the graph, we need to draw some horizontal and vertical lines.

Draw ***horizontal*** lines from the left side of the graph at every ***nosecone length*** on the ordinate, over to the right hand side of the graph.

Draw a **vertical** line from the bottom of the graph at each abscissa value to the top of the graph.

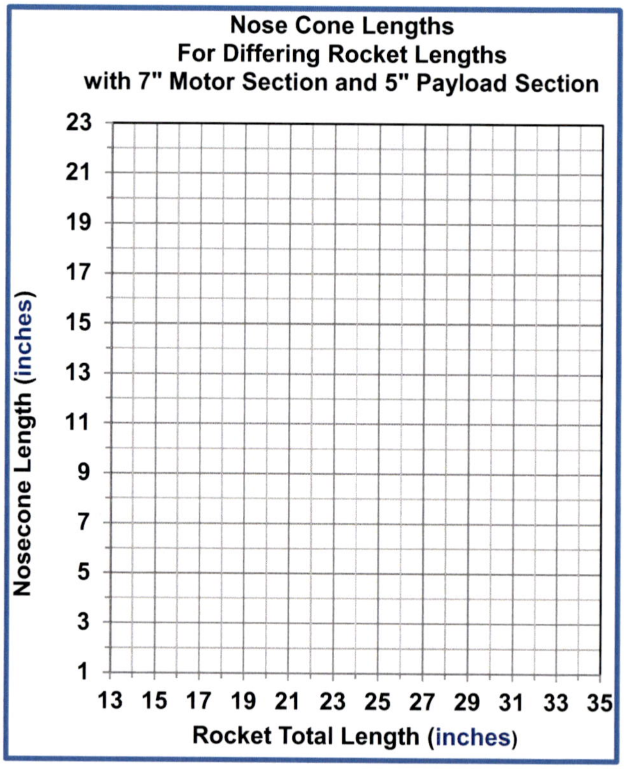


**Figure 18: (Step 8 of 10) Add vertical and horizontal lines**

Once you have the vertical and horizontal lines completed, we can put our data points at the intersection of these lines.

**Okay, let's put our data on the graph!**

### Step 9:

In this step, we will apply our data points to our graph. Soon we will be able to predict what nosecone length to use for ***any*** rocket length between 15 and 33 inches.

When you're done with Step 9, you should have 10 data points on the graph, one for each rocket we are designing.

**Okay, let's go on to Step 10 and finish our graph so we can use it!**

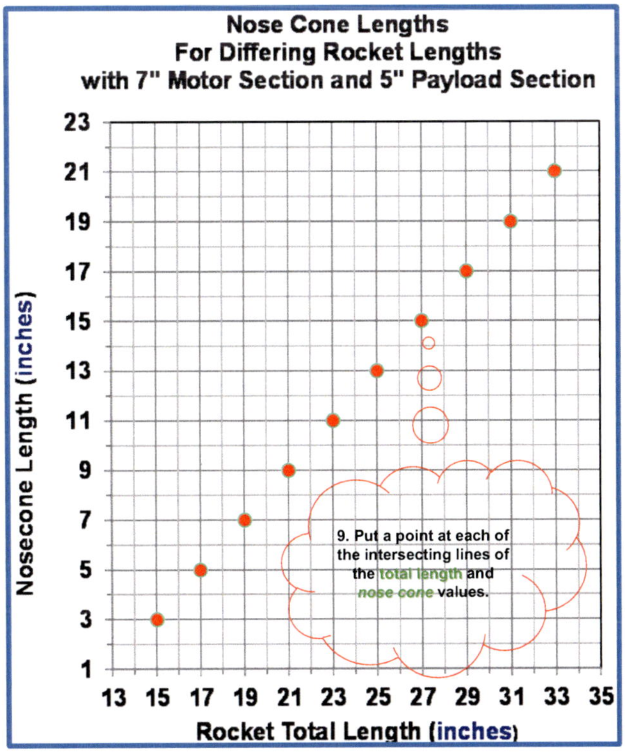


**Figure 19: (Step 9 of 10) Placing data points on the graph**

### Step 10:

Draw a line through the data points.

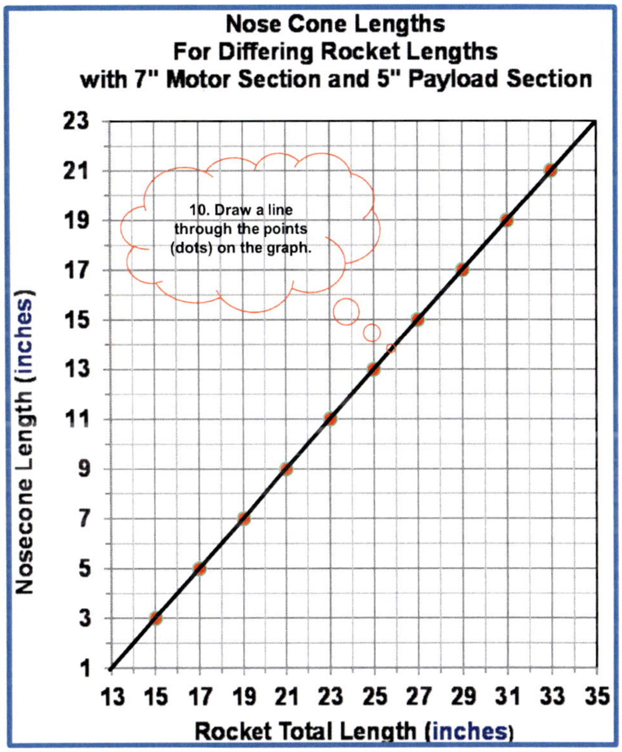


**Figure 20: (Step 10 of 10) Draw line through data points to complete graph**

Great! Now we have a graph that can be used to predict any nosecone length for any rocket size between 15 inches and 33 inches long!

Look at the graph in ***Figure 20***. See if you can tell what the nosecone length would be for a 24 inch long rocket

Look at the abscissa and find the tic mark at 24 inches.

Draw a vertical line up to our line that passes through the data points we just created.

Draw a horizontal line from there to the ordinate axis and read the value.

That's right! Our nosecone should be 12 inches long. Cool huh!

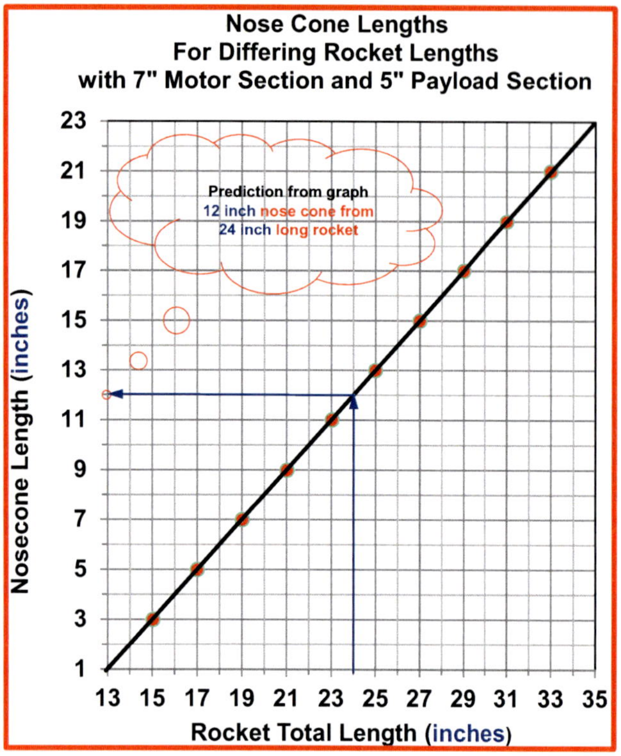


**Figure 21: Prediction from graph for 24-inch-long rocket**

Now is a great time to ask your teacher about what is called "**the equation of a line**".

That's right; we can write another mathematical expression that defines our picture of the graph!

### 4.6 Create a lab to verify or deny the work in 1-5

1. Draw a picture of the problem
2. Label the picture
3. Write a mathematical expression that defines the picture
4. Solve for the unknowns
5. Calculate the results (by hand, by calculator, or by computer application software)
6. **Create a lab to verify or deny the work in 1-5**
7. Test the results of 1-5 in the lab

In LAB 1, we created a lab to test our work. In this lab, we will pick any sized rocket from the graph and repeat the steps used in LAB 1.

**We will draw each section of the rocket on a NEW piece of paper using the dimensions from the drawing and from the graph.** Then we will tape all the pieces together. Once they are all taped together, the total length of our rocket should come out to $L_T$ inches long. Remember, $L_T$, and $L_N$ will be different for each student and will depend on the rocket length chosen.

Below is the procedure for LAB 2:

1. Cut a piece of paper 7 inches long and 2 inches wide for the motor section (*this has not changed from LAB 1*).
2. Cut another piece of paper 5 inches long and 2 inches wide for the payload section (*this has not changed from LAB 1*).
3. Cut another piece of paper for the nosecone. The length you choose here will be determined by how long you want your rocket to be, $L_T$. Pick your length from the abscissa and determine the length of your nosecone, $L_N$. Make sure the nosecone section is 2 inches wide.

You can use any length on the abscissa. You can pick a point on the abscissa between the numbers too! For example, if you want a rocket that is 30 inches long, just start a vertical line between the numbers 29, and 31 on the abscissa.

4. Once you have cut out the rectangle for the nosecone, fold it in half and cut a curved section out of this piece so it looks like a nosecone.
5. Make two fins that look like the one in our drawing.
6. Make a flame to come out the motor section.
7. Tape all the pieces together so it looks like our drawing.

**How long is your rocket going to be? Don't forget your units!**

______________________________

______________________________

**From the graph, how long is your nosecone going to be? Don't forget your units!**

______________________________

______________________________

## 4.7 Test the results of 1-5 in the lab.

1. Draw a picture of the problem
2. Label the picture
3. Write a mathematical expression that defines the picture
4. Solve for the unknowns
5. Calculate the results (by hand, by calculator, or by computer application software)
6. Create a lab to verify or deny the work in 1-5
7. **Test the results of 1-5 in the lab**

Once you have taped all the pieces together, **measure the TOTAL LENGTH** of the rocket to see if it is as long as it is supposed to be. If the overall length of your rocket is correct, the nosecone length must also be correct!

## 5 Results (Compare math with real life application)

You probably found out that the ***graphical*** prediction we used to solve for the length of the nosecone was very close to your finished cutout.

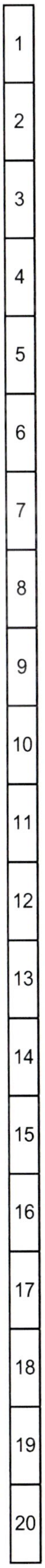


Even if you may not have gotten the exact total length, your measured value was probably very, very close to the total length. The nosecone length you selected from the graph was also very close to what the graph predicted. You may have seen slight variations in the lengths. This is due to the same errors we saw in LAB 1. When slight variations happen like this, we call it **error**. Here is a review of the errors discussed in LAB 1, along with some new causes.

Some cause of **error** might be:

- Your scissors may have been dull and you couldn't follow the outline of your drawing perfectly.
- Or, maybe when you taped the pieces together they shifted somewhat.
- There may have been some error in the vertical and horizontal lines. This may give you a nosecone length that was slightly off.
- There may have been other reasons for the error. Remember that, even when we use the most precise tools, we still might experience some precision and accuracy errors.

## 6 Conclusions (What did we learn?)

- In this lab, we used the same mathematical expression we created in LAB 1.
- Remember that we created that mathematical expression from a picture.
- In this lab, we used the same

mathematical expression to create a table and a graph.

- Then we used the graph to predict different nosecone lengths for different sized rockets without looking at the mathematical expression.
- The graph was created using 10 Steps.
- Can you see why, once we create a mathematical expression, it's very convenient to show it in a table or in a graph?
- As in every lab, go back to the beginning of this lab and read the abstract (learning outcome) again.

Does it make more sense now?

***Where have you seen other tables and graphs? Have you seen them in some of the books you're using at school?***

____________________________________

____________________________________

____________________________________

____________________________________

It's a good idea to show the mathematical expression on the graph so you can refer back to the mathematical expression used to create the graph. The graph in ***Figure 22*** shows the completed graph with the mathematical expression included.

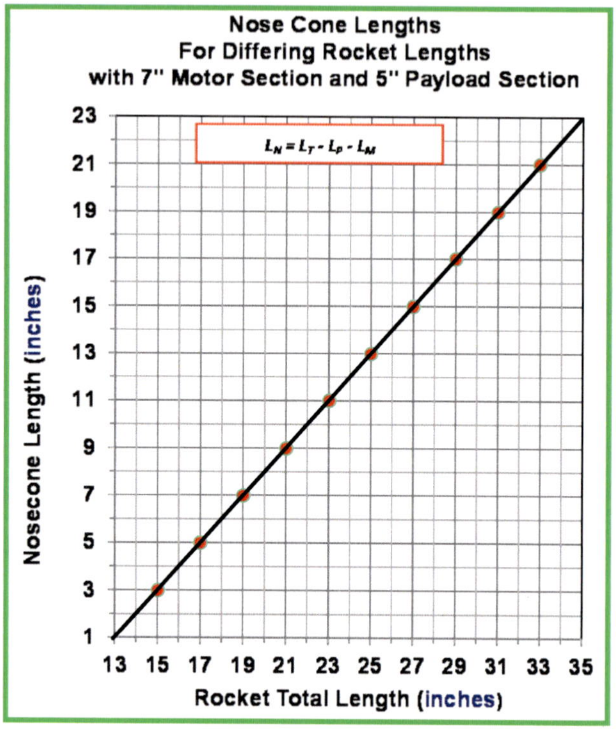


**Figure 22: Graph with mathematical expression included**

## 7 References and additional reading

**7.1** **The Handbook of Model Rocketry by G. Harry Stine**

**7.2** **The Parachute Manual by Dan Poynter**

**7.3** **National Association of Rocketry, www.nar.org**

**7.4** **Email: nar-hq@nar.org**

**7.5** **http://www.nasa.gov**

**7.6** **http://www.Spacex.com**

# 8 INDEX

# LAB 3: CALCULATING THE STABILITY OF A ROCKET

# LAB 3: CALCULATING THE STABILITY OF A ROCKET

4th - 6th Grade Learning Standards Applied to High Powered Rocket Flight and Descent

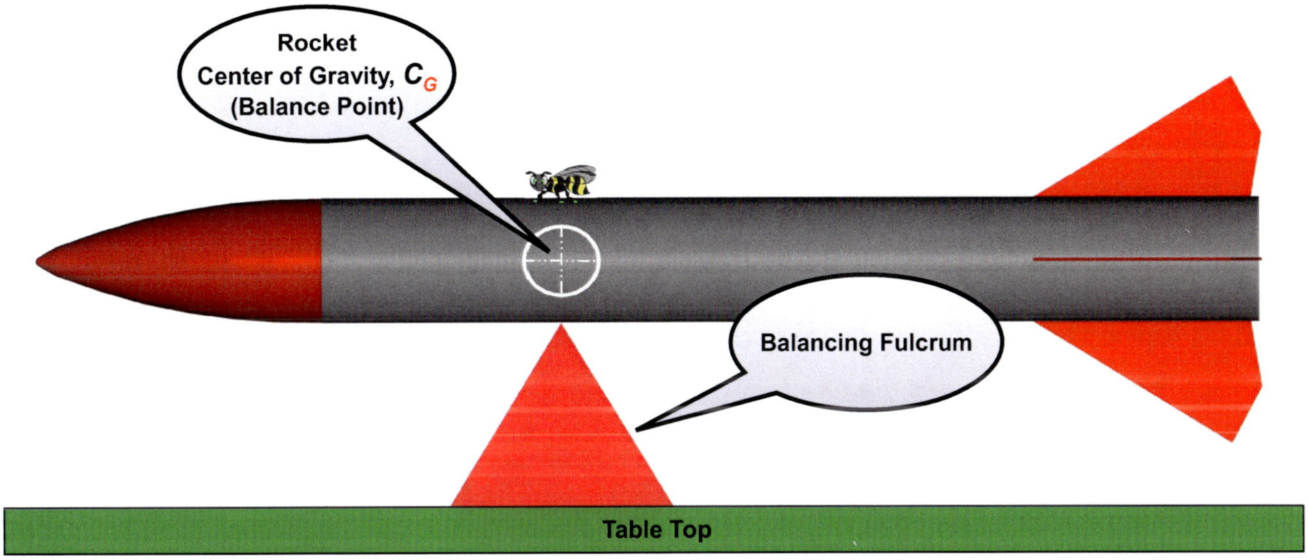


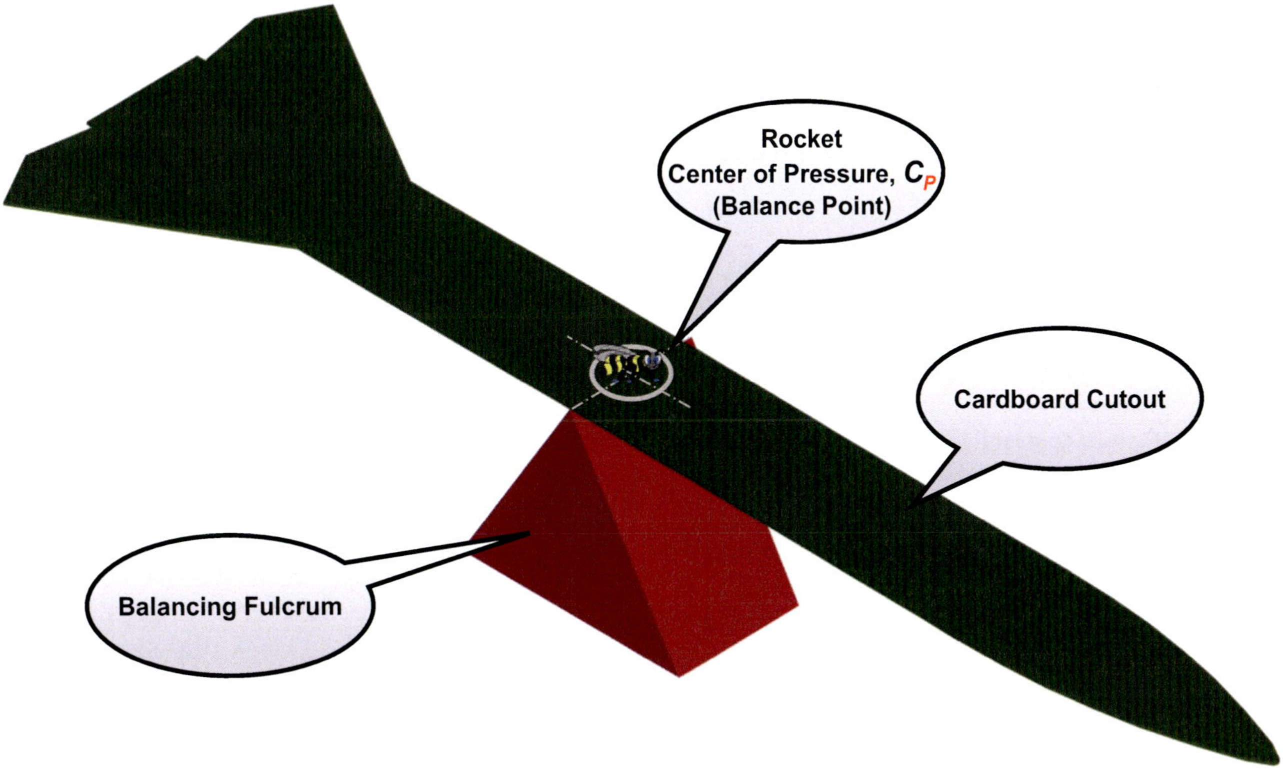

## Table of Contents



## List of Tables and Figures

**Figure 1: Brave Bees on for the ride**

## 1 Abstract (What are we going to learn?)

In order for a rocket to fly in a straight path, it must be "***STABLE***". For a rocket to fly a stable flight, it must be designed so that its *center of gravity* **leads** its *center of pressure*. The center of gravity for any object is its balance point. The center of pressure is the balance point of the air pressure.

Without this relationship, where the center of gravity leads the center of pressure, a rocket will cartwheel out of control and do a lot of crazy things in flight. We call this condition ***"UNSTABLE"***. If a rocket is unstable, it is very dangerous and could really hurt someone not to mention cause the rocket to crash.

What does "lead" mean? Well, picture two brave bees riding on a rocket. As the rocket flies, one bee, named "**Gravity**", is higher up on the rocket, *closer* to the nosecone. The other bee, named "**Pressure**", is riding on the lower part of the rocket *farther* away from the nosecone. If these two bees were considered to be in a race while the rocket was in flight, the bee near the top would always be in the lead. That's why we say, **"The Center of Gravity leads the Center of Pressure".**

We are going to learn how to check out a rocket to confirm whether or not it is stable (whether or not it will fly in a straight and controlled path).

You can remember the relationship of the center of gravity, $C_G$, and the center of pressure, $C_P$, by saying the following:

***"Gravity leading pressure… makes flight a pleasure…***

In this lab, we are going to discuss the "***rule of thumb***" for checking the relationship between the $C_G$ and $C_P$. To do this, we will carry out the steps necessary to verify that a rocket is indeed stable and ready for a safe flight.

## 2 Glossary of Terms (New words and symbols we need to learn)

**FROM LAB2**:

**Abscissa axis**: This is a word used to describe the axis on a graph that is normally horizontal. The numbers on this axis determine the numbers on the ordinate axis.

**Ordinate axis**: This is a word used to describe the axis on a graph that is normally vertical. The numbers on the abscissa axis change the numbers on this axis.

**NEW TERMS**:

**Lead:** When some event occurs first, it is said to lead, just as in a race between two runners.

**Lag:** When some event occurs after another event, that event is said to lag the other.

**NAR:** National Association of Rocketry: The National Association of Rocketry is the governing body for model rocketry enthusiasts. The primary goals of the organization are to ensure safety and education of model rocketry to model rocket enthusiasts.

**Stable:** When a rocket flies in a straight predictable path it is considered stable. This condition exists when the center of gravity **leads** the center of pressure.

**Unstable:** When a rocket flies in an uncontrolled and unpredictable path, it is considered unstable. This condition exists when the center of gravity **lags** the center of pressure.

**Yikes:** A term Hoosiers (people from Indiana) use when something goes terribly wrong. ***Oh yikes!***

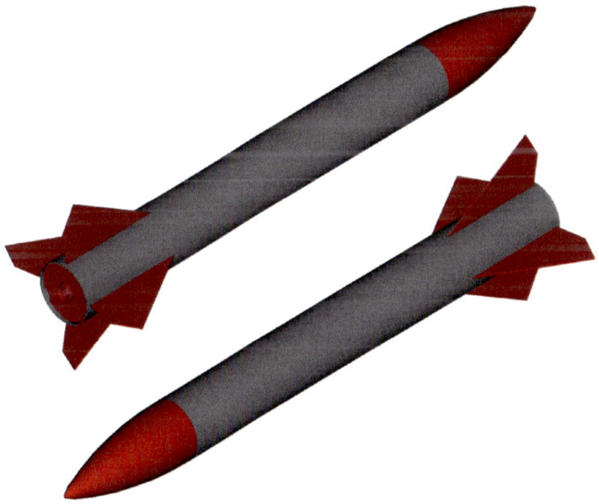

**Figure 2: Three-dimensional illustration of a rocket**

### Other Symbols used:

**Φ** This is a symbol used to represent the **diameter** of our rocket (in inches).

$C_G$ **Center of Gravity:** This is the location on the rocket where it is balanced when placed on a fulcrum. The $C_G$ is a distance (in inches) usually measured from the nosecone tip to the location on the rocket where the rocket will balance on a fulcrum. The center of gravity of a person is just below the belly button.

$C_G$ *(in)*

**$C_P$ Center of Pressure:** This is the location where the forces due to air bombarding a surface are balanced. The $C_P$ is a distance (in inches) that can be measured on a cardboard rocket cut-out. It is measured from the nosecone tip to the location where the cut-out will balance on a fulcrum.

$$C_P\,(in)$$

***d*** This is a symbol used to represent the **distance between the center of gravity and the center of pressure** of a rocket (in inches).

$$d = C_P\,(in) - C_G\,(in)$$

Δ This is a symbol you can use to replace the words "**distance or difference between**", or just plain "**delta**". For example, we can write:

$$d = \Delta C = C_P\,(in) - C_G\,(in)$$

**=** The equal sign means that two quantities have the same value. This is important when a cake is being cut and you want your = share.

**<** This symbol replaces the phrase "**less than**". It would be unfortunate if your piece of cake were < your friend's.

**>** This symbol replaces the phrase "**greater than**". It would be great if your piece of cake were > your friend's.

**≤** This symbol replaces the phrase "**less than OR equal**". This means that a quantity can be less than or equal to another quantity, but cannot be larger. Is it a good thing that your friend's cake is ≤ to your piece of cake?

**≥** This symbol replaces the phrase "**greater than OR equal**". This means that a quantity can be greater than or equal to another quantity, but cannot be smaller. Is it a good thing that your friend's cake is ≥ to your piece of cake?

**"In a nutshell":** This means to wrap up something complicated into a summary.

"**In a beehive**": This also means to wrap up something complicated into a summary but doesn't involve squirrels.

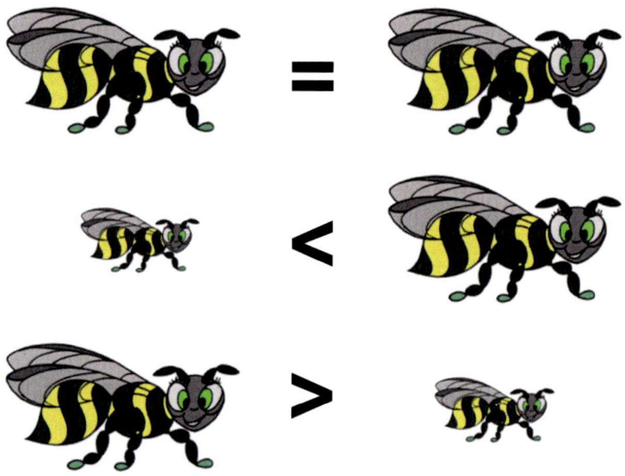


**Figure 3: Illustration of the "equal" and "inequality" symbols**

## 3 Introduction and Purpose (Why are we doing this?)

Later, we will be building a rocket to fly, but before we fly our rocket, we have to check to see if it is stable. Model rockets travel at very high rates of speed. It is not unusual for a rocket to travel at speeds in excess of **140 miles per hour in a little over a second** just after liftoff.

If a rocket is not designed to fly in a straight vertical path, serious injury can occur to anyone within hundreds of feet of the launch pad.

If a rocket is STABLE, we can be sure that its path will be straight. The purpose of this lab is to verify that our rocket is indeed stable.

There are some illustrations of how well a rocket will fly in ***Figure 8***. The illustrations are based on the positions of the center of gravity and center of pressure.

## 4 Procedure: (How do we do it?)

1. **Draw a picture of the problem**
2. **Label the picture**
3. **Write a mathematical expression that defines the picture**
4. **Solve for the unknowns**
5. **Calculate the results (by hand, by calculator, or by computer application software)**
6. **Create a lab to verify or deny the work in 1-5**
7. **Test the results of 1-5 in the lab**

**Figure 4: Seven steps to solving a problem**

As we have done in all of our previous labs, we are going to follow the seven procedures to solve our problem.

As you will see, the steps for solving this lab are identical to the steps in LAB 1 and LAB 2. If you have LAB 1 or LAB 2 handy, you may want to refer to them from time to time as you read this lab.

### 4.1 Draw a picture of the problem

1. Draw a picture of the problem
2. Label the picture
3. Write a mathematical expression that defines the picture
4. Solve for the unknowns
5. Calculate the results (by hand, by calculator, or by computer application software)
6. Create a lab to verify or deny the work in 1-5
7. Test the results of 1-5 in the lab

In the first lab, we drew a picture of the rocket, so the first step of our lab is complete. If you don't have the picture, just draw one similar to the rocket in ***Figure 5***. Let's go ahead and do that.

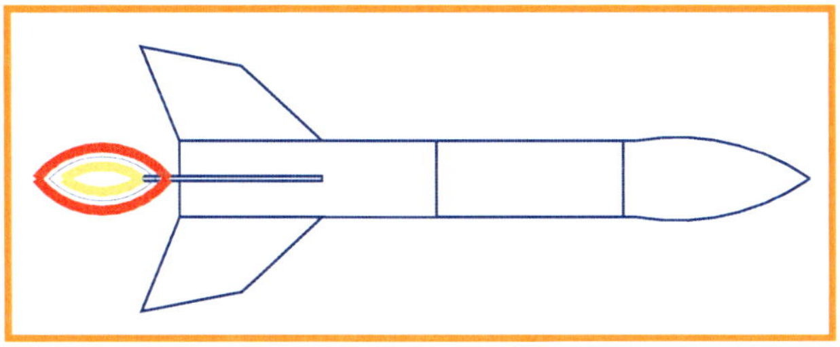

**Figure 5: Completed rocket with cool flames**

**Remember, the picture is the most important thing to do.** Once we have the picture, we have something to refer to as we continue to solve the problem.

After you have drawn your picture, go ahead to the next step and label it like the one illustrated in ***Figure 7***.

### 4.2 Label the picture

1. Draw a picture of the problem
2. Label the picture
3. Write a mathematical expression that defines the picture
4. Solve for the unknowns
5. Calculate the results (by hand, by calculator, or by computer application software)
6. Create a lab to verify or deny the work in 1-5
7. Test the results of 1-5 in the lab

Notice that the dimensions (distances) are measured from the nosecone tip. Since both the *center of gravity* and the *center of pressure* have been measured from this point, we can write a mathematical expression that defines our picture.

***Remember**, every time you read a math or science book, there will be a picture that represents the problem.*

If you don't see a picture, read the words to see if you can draw a picture about the problem. Once you've drawn it, go ahead and label it properly.

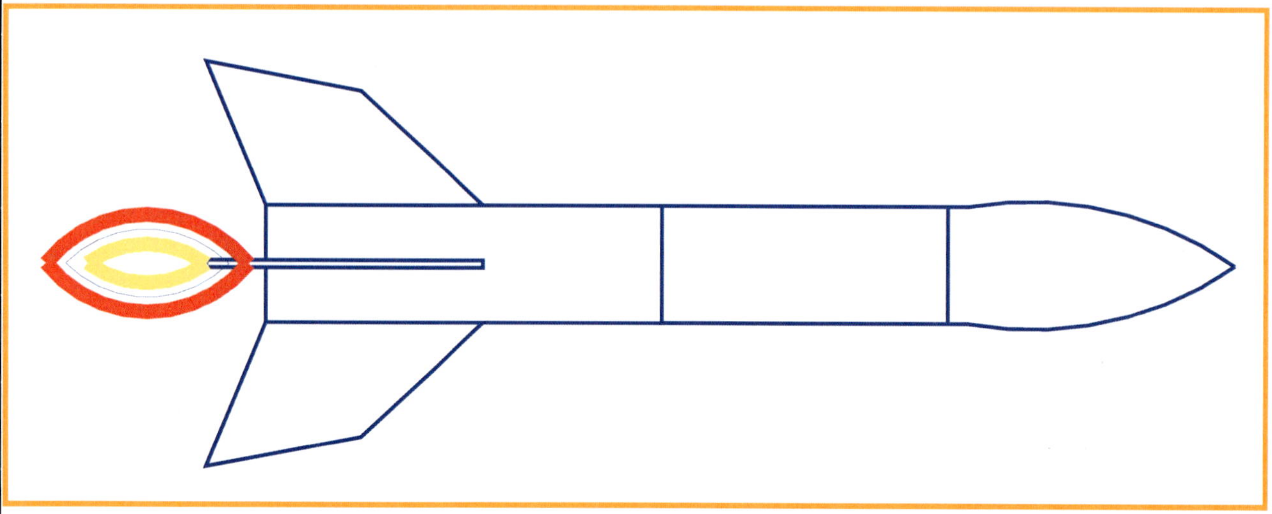

**Figure 6: Original picture of our rocket**

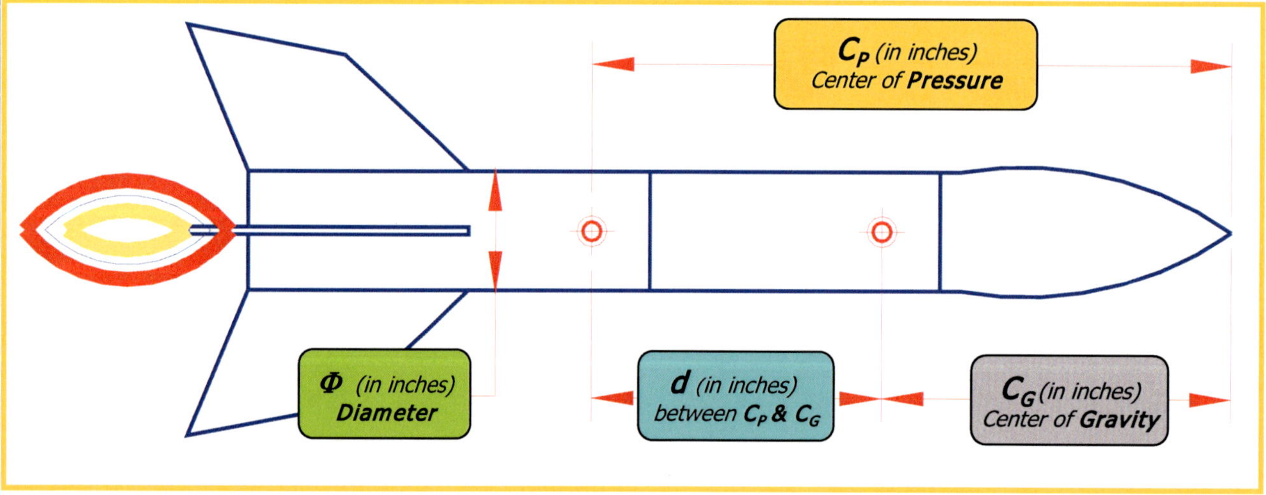


**Figure 7: Rocket labeled with center of gravity and center of pressure**

### 4.3 Write a mathematical expression that defines the picture

1. Draw a picture of the problem
2. Label the picture
3. **Write a mathematical expression that defines the picture**
4. Solve for the unknowns
5. Calculate the results (by hand, by calculator, or by computer application software)
6. Create a lab to verify or deny the work in 1-5
7. Test the results of 1-5 in the lab

Look at the picture of the rocket you just labeled. Can you see why we can write the following sentence?

***"The distance from the nosecone tip to the center of gravity, $C_G$ IS LESS THAN the distance from the nosecone tip to the center of pressure, $C_P$ (both are in inches)".***

Using the less than symbol, it is written this way:

$$C_G \text{ (inches)} < C_P \text{ (inches)}$$

The symbol < is the symbol we use to replace the words, "***less than***" in a mathematical sentence. It may look somewhat strange at first but if you just remember that every symbol represents words out of a sentence, it's not so bad. An advantage to writing the sentence this way (*with symbols*) is that it takes up less space on a piece of paper and can be written quickly. Also, the *symbols are the same anywhere in the world*. So you could say, that you speak a little of every language on the planet. The one thing to keep in mind, though, is that without the picture, we would not be able to write this mathematical expression.

Look again at the picture of the rocket you just labeled, ***Figure 7***. Can you see why we can write the following sentence?

***"The distance, from the nosecone tip to the center of pressure, $C_P$, minus the distance from the nosecone tip to the center of gravity, $C_G$, equals the distance between them, d".***

Written in symbols, it looks like this:

$$d = C_P \text{ (inches)} - C_G \text{ (inches)}$$

We can also say:

***"The distance, d, between the $C_G$ and the $C_P$ should be somewhere between just a little bit over zero inches, but less than or equal to the rocket diameter in inches".***

This is what is referred to as a "***rule of thumb***". This is really meant to be a guideline more than a rule. This sentence is written this way:

$$0 < (C_P - C_G) \leq \Phi$$

Since ***d*** is equal to ($C_P$ - $C_G$), we can stuff that into our mathematical expression so it looks like this:

$$0 < d \leq \Phi$$

By the way, our mathematical expression above says the same thing as **all** of the written sentences so far, but we've combined them into one!

**Notice how much shorter our symbol sentence is than all of the combined word sentences.**

***"The distance from the nosecone to the center of gravity, $C_G$ (in inches), is less than the distance from the nosecone to the center of pressure, $C_P$ (in inches)".***

$$C_G \text{ (in)} < C_P \text{ (in)}$$

***"The distance from the nosecone to the center of pressure, $C_P$ (in inches), minus the distance from the nosecone to the center of gravity, $C_P$ (in inches), equals the distance between them".***

$$d = C_P \text{ (in)} - C_G \text{ (in)}$$

***"The distance, d, between the $C_G$ and the $C_P$ should be somewhere between just a little bit over zero inches, but less than or equal to the rocket diameter in inches".***

$$0 < d \leq \Phi$$

So, "*in a beehive*", this tells us is that, if you measure the center of gravity and mark the rocket in that location, and measure the center of pressure and mark its location on the rocket, your measurement between the marks should be somewhere between just over zero and the diameter of the rocket. The picture below shows approximate paths of rockets designed with different "*d*" values. All but one of the illustrations follows the rule: *$C_G$ leads $C_P$. The distance, d, between the $C_G$ and $C_P$ should be between zero and the diameter*. **Which rocket did not follow the rule**?

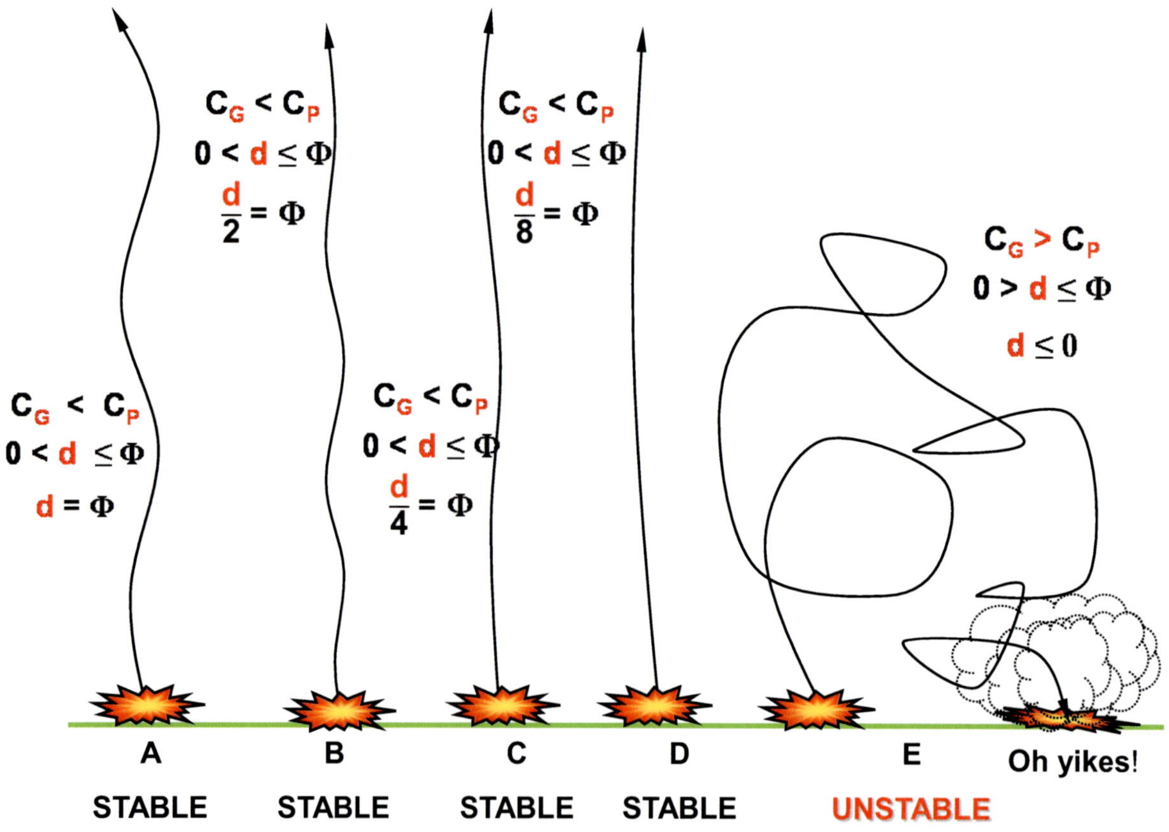


**Figure 8: Illustration of different flight paths**

**If the center of gravity and the center of pressure stayed in the same place on our picture, but instead, we decided to measure from the motor end of the rocket, what would happen to our mathematical sentence?**

___

___

**Would our mathematical expression change? If so; why? Can you write it?**

___

___

### 4.4 Solve for the unknowns

1. Draw a picture of the problem
2. Label the picture
3. Write a mathematical expression that defines the picture
4. **Solve for the unknowns**
5. Calculate the results (by hand, by calculator, or by computer application software)
6. Create a lab to verify or deny the work in 1-5
7. Test the results of 1-5 in the lab

The next step is solving for the unknowns. For our mathematical expression, there is really nothing to solve.

We have already written the mathematical expression in a useful form. Since we are checking to make sure that the distance from the nosecone to the center of gravity is less than the distance from the nosecone to the center of pressure, our mathematical expression is ready to use!

Let's put our new knowledge to work. Let's imagine for a minute that we have to help out some rocket designers that don't really understand how to make a rocket stable. They've come to us for some help. They've told us they built a rocket with a 1 inch diameter and they know the $C_P$ is located 10 inches from the nosecone tip. The $C_P$ and the diameter of the rocket will not change.

They also found the $C_G$ of the rocket fully fueled and wrote down the $C_G$ (without a payload).

Then they placed their payload in different locations along the rocket and wrote down the $C_G$ for each location. (*If we add a payload to the rocket, the $C_G$ WILL change. Where the $C_G$ ends up is determined by WHERE the extra weight is added*).

They were nice enough to give us a table of their findings and want us to tell them which designs are STABLE or UNSTABLE See ***Figure 11***.

In other words, we need to find out where it's safe to put the payload (parachute, camera, egg, or any or type of payload) inside the rocket and still have a stable rocket.

Here is something interesting to think about. The $C_G$ moves closer to the nosecone as fuel is consumed during the ascent stage of the rocket flight. We'll talk more about that later but if you want to look ahead you'll see this in ***Figures 21 & 22***. WHAT'S IMPORTANT IS THAT YOU WANT TO MAKE SURE YOUR ROCKET HAS THE SAME FUEL WEIGHT THAT IT WOULD AT LIFTOFF BEFORE YOU LOCATE THE $C_G$.

Before we begin filling out our table, take a look at the illustrations in ***Figures 9 & 10***. On the next page you'll find a review of the mathematical expressions. They are also written in the top of the table in ***Figure 11***.

If we follow these three rules, and everything else goes right, we will have a beautiful flight.

**We want to use our "*rule of thumb*". It states:**

$$C_G < C_P \;\; \underline{\text{and}} \;\; d > 0 \;\; \underline{\text{and}} \;\; d \le \Phi$$

*In words, any distance, **d**, between the $C_P$ and the $C_G$ is a good value as long as:*

1. ***The distance from the nosecone tip to $C_G$ is less than the distance from the nosecone tip to the $C_P$.***

$$C_G < C_P$$

2. ***"d" is greater than zero.***

$$d > 0$$

3. ***"d" less than or equal to the diameter of the rocket.***

$$d \le \Phi$$

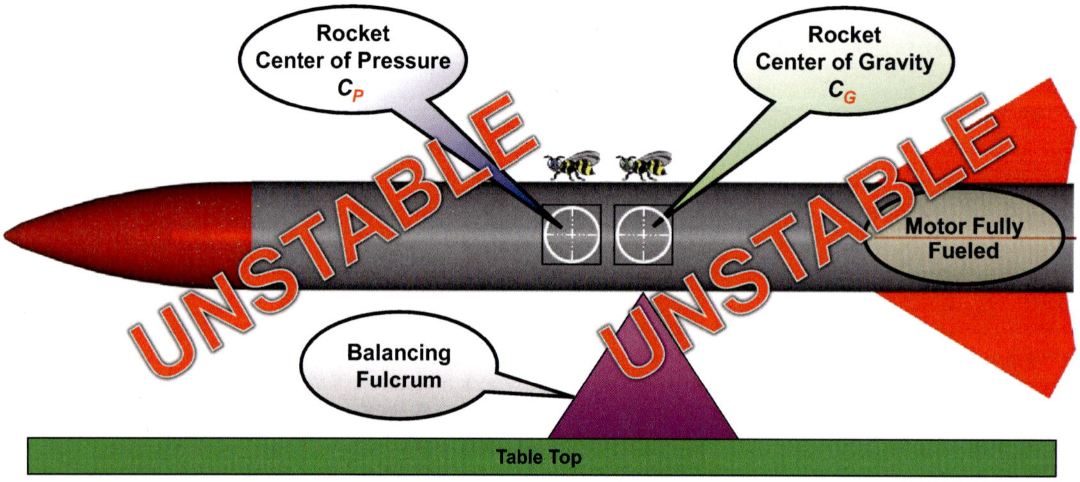


**Figure 9: Rocket Center of Gravity BEFORE Adding Payload**

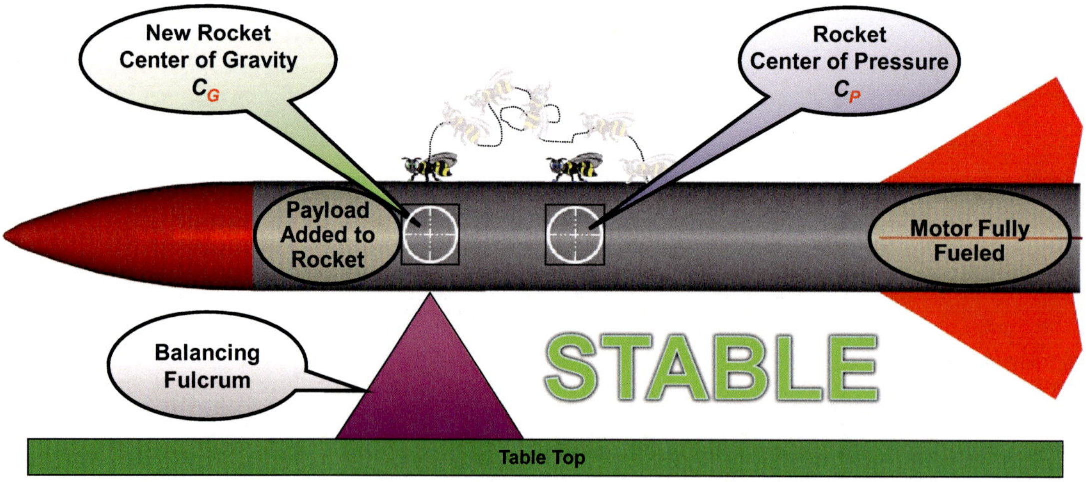


**Figure 10: Rocket Center of Gravity AFTER Adding Payload**

| ROCKET # | Diameter of rocket inches $\phi$ Given to us | Distance from tip to $C_P$ inches $C_P$ Given to us | Distance from tip to $C_G$ inches $C_G$ Given to us | Distance from $C_P$ to $C_G$ inches $d = C_P - C_G$ We calculate | IS $C_G < C_P$ YES / NO | IS $0 < d \le \Phi$, $0 < (C_P - C_G) \le \Phi$: $d > 0$ YES / NO | IS $0 < d \le \Phi$, $0 < (C_P - C_G) \le \Phi$: $d \le \Phi$ YES / NO | IS THE ROCKET STABLE? YES / NO |
|---|---|---|---|---|---|---|---|---|
| 1 | 1 | 10 | 8 | | | | | |
| 2 | 1 | 10 | 8 1/4 | | | | | |
| 3 | 1 | 10 | 8 1/2 | | | | | |
| 4 | 1 | 10 | 8 3/4 | | | | | |
| 5 | 1 | 10 | 9 | | | | | |
| 6 | 1 | 10 | 9 1/4 | | | | | |
| 7 | 1 | 10 | 9 1/2 | | | | | |
| 8 | 1 | 10 | 9 3/4 | | | | | |
| 9 | 1 | 10 | 10 | | | | | |
| 10 | 1 | 10 | 10 1/4 | | | | | |
| 11 | 1 | 10 | 10 1/2 | | | | | |
| 12 | 1 | 10 | 10 3/4 | | | | | |
| 13 | 1 | 10 | 11 | | | | | |

**Figure 11: Table using the rule of thumb**

## 4.5 Calculate the results (by hand, by calculator, or by computer application software)

1. Draw a picture of the problem
2. Label the picture
3. Write a mathematical expression that defines the picture
4. Solve for the unknowns
5. **Calculate the results (by hand, by calculator, or by computer application software)**
6. Create a lab to verify or deny the work in 1-5
7. Test the results of 1-5 in the lab

Remember, we have been given a rocket with a 1-inch diameter. We have been told that the center of pressure, $C_P$, is 10 inches from the nosecone tip.

Let's look at a table of data for a rocket that's being designed in ***Figure 11***. It shows us *different* values for the center of gravity, $C_G$ based on adding a payload at different locations inside the rocket.

We have to fill in the **five right hand columns** based on the information from the **left three columns.**

We need to calculate, $C_P$ - $C_G$, and then determine whether or not the rocket will fly properly based on the dimensions given to us.

If we find that the values agree with our rule of thumb, we will know that the rocket will fly properly. It has been left up to us to notify the designers which rocket will fly, and which ones may cause us to say "***Oh Yikes***" after a launch.

The following five illustrations show you how to fill out the table. Once you have done the first row, continue down the rows, filling out each row just as you did the first row.

## INSTRUCTIONS FOR FILLING OUT OUR ROCKET STABILITY TABLE

### STEP 1 of 5:

First we calculate the distance, *d*, between the Center of Gravity, $C_G$, and the Center of Pressure, $C_P$, and place the answer in the $d = C_P - C_G$ column in the table.

$$d = C_P - C_G,$$

| ROCKET # | Diameter of rocket inches | Distance from tip to $C_P$ inches | Distance from tip to $C_G$ inches | Distance from $C_P$ to $C_G$ inches | IS | IS $0 < d \leq \Phi$ $0 < (C_P - C_G) \leq \Phi$ | | IS THE ROCKET STABLE? |
|---|---|---|---|---|---|---|---|---|
| | $\phi$ | $C_P$ | $C_G$ | $d = C_P - C_G$ | $C_G < C_P$ | $d > 0$ | $d \leq \Phi$ | |
| | Given to us | Given to us | Given to us | We calculate | YES / NO | YES / NO | YES / NO | YES / NO |
| 1 | 1 | 10 | 8 | 2 = 10 - 8 | | | | |
| 2 | 1 | 10 | 8 1/4 | | | | | |
| 3 | 1 | 10 | 8 1/2 | | | | | |
| 4 | 1 | 10 | 8 3/4 | | | | | |
| 5 | 1 | 10 | 9 | | | | | |
| 6 | 1 | 10 | 9 1/4 | | | | | |
| 7 | 1 | 10 | 9 1/2 | | | | | |
| 8 | 1 | 10 | 9 3/4 | | | | | |
| 9 | 1 | 10 | 10 | | | | | |
| 10 | 1 | 10 | 10 1/4 | | | | | |
| 11 | 1 | 10 | 10 1/2 | | | | | |
| 12 | 1 | 10 | 10 3/4 | | | | | |
| 13 | 1 | 10 | 11 | | | | | |

**Figure 12: The distance between the Center of Gravity and the Center of Pressure**

## INSTRUCTIONS FOR FILLING OUT OUR ROCKET STABILITY TABLE:

### STEP 2 of 5:

In the next column, we check to see if center of gravity, $C_G$, is less than, <, the center of pressure, $C_P$.

Since the center of gravity measured from the tip of the rocket is ***8*** inches, and the center of pressure measured from the tip of the rocket was **10** inches, we will mark this column with a *YES*. **In other words:**

$$C_G < C_P$$

**Note**:

**If the center of gravity, $C_G$, is NOT less than or equal to the center of pressure, $C_P$, the rocket is *UNSTABLE*!**

| ROCKET # | Diameter of rocket inches $\phi$ Given to us | Distance from tip to $C_P$ inches $C_P$ Given to us | Distance from tip to $C_G$ inches $C_G$ Given to us | Distance from $C_P$ to $C_G$ inches $d = C_P - C_G$ We calculate | IS $C_G < C_P$ YES / NO | IS $0 < d \leq \Phi$ $0 < (C_P - C_G) \leq \Phi$ $d > 0$ YES / NO | $d \leq \Phi$ YES / NO | IS THE ROCKET STABLE? YES / NO |
|---|---|---|---|---|---|---|---|---|
| 1 | 1 | 10 | 8 | 2 = 10 - 8 | YES | | | |
| 2 | 1 | [illegible] | [illegible] | | | | | |
| 3 | 1 | [illegible] | [illegible] | | | | | |
| 4 | 1 | [illegible] | [illegible] | | | | | |
| 5 | 1 | [illegible] | [illegible] | | | | | |
| 6 | 1 | 10 | 9 1/4 | | | | | |
| 7 | 1 | 10 | 9 1/2 | | | | | |
| 8 | 1 | 10 | 9 3/4 | | | | | |
| 9 | 1 | 10 | 10 | | | | | |
| 10 | 1 | 10 | 10 1/4 | | | | | |
| 11 | 1 | 10 | 10 1/2 | | | | | |
| 12 | 1 | 10 | 10 3/4 | | | | | |
| 13 | 1 | 10 | 11 | | | | | |

**Figure 13: Center of Gravity less than, <, Center of Pressure distance check**

## INSTRUCTIONS FOR FILLING OUT OUR ROCKET STABILITY TABLE

### STEP 3 of 5:

Next, we will check to see if the distance, ***d***, is greater than *zero*. In Step 1, we calculated that distance to be ***2** inches*. Since the distance 2, is greater than *0*, we will mark the column with a YES. In Step 2 we said that $C_G < C_P$. If $C_G < C_P$, ***d*** has to be greater than ***0***!

***d > 0, so we fill in a YES!***

**Note:**

**If the distance, *d*, between the center of gravity, $C_G$, and the center of pressure, $C_P$, is NOT greater than, > , zero, the rocket is UNSTABLE!**

| ROCKET # | Diameter of rocket inches $\phi$ Given to us | Distance from tip to $C_P$ inches $C_P$ Given to us | Distance from tip to $C_G$ inches $C_G$ Given to us | Distance from $C_P$ to $C_G$ inches $d = C_P - C_G$ We calculate | IS $C_G < C_P$ YES / NO | IS $0 < d \leq \Phi$, $0 < (C_P - C_G) \leq \Phi$: $d > 0$ YES / NO | $d \leq \Phi$ YES / NO | IS THE ROCKET STABLE? YES / NO |
|---|---|---|---|---|---|---|---|---|
| 1 | 1 | 10 | 8 | 2 = 10 - 8 | YES | YES | | |
| 2 | 1 | 10 | 8 1/4 | | | | | |
| 3 | 1 | 10 | 8 1/2 | | | | | |
| 4 | 1 | 10 | 8 3/4 | | | | | |
| 5 | 1 | 10 | 9 | | | | | |
| 6 | 1 | 10 | 9 1/4 | | | | | |
| 7 | 1 | 10 | 9 1/2 | | | | | |
| 8 | 1 | 10 | 9 3/4 | | | | | |
| 9 | 1 | 10 | 10 | | | | | |
| 10 | 1 | 10 | 10 1/4 | | | | | |
| 11 | 1 | 10 | 10 1/2 | | | | | |
| 12 | 1 | 10 | 10 3/4 | | | | | |
| 13 | 1 | 10 | 11 | | | | | |

**Figure 14: Distance between the Center of Gravity and the Center of Pressure Zero check**

## INSTRUCTIONS FOR FILLING OUT OUR ROCKET STABILITY TABLE:

### STEP 4 of 5:

Next, we check to see if the distance, ***d,*** between the Center of Gravity $C_G$ and the Center of Pressure, $C_P$, is less than or equal, ≤, to the diameter, $\Phi$, of our rocket, ($d \le \Phi$).

Since our distance between the center of gravity and the center of pressure was calculated to be ***2 inches***, and the diameter of our rocket is ***1 inch***, our answer is NO. In other words:

***d is NOT ≤ Φ so we fill in a NO!***

**Note:**

**If the distance, *d*, between our center of pressure, $C_P$, and our center of gravity, $C_G$, is NOT less than or equal, ≤, to the diameter, $\Phi$, the rocket may still be *STABLE*, but this may not be our best design. Remember our rule of thumb from *Section 4.3*.**

| ROCKET # | Diameter of rocket inches $\phi$ Given to us | Distance from tip to $C_P$ inches $C_P$ Given to us | Distance from tip to $C_G$ inches $C_G$ Given to us | Distance from $C_P$ to $C_G$ inches $d = C_P - C_G$ We calculate | IS $C_G < C_P$ YES / NO | IS $0 < d \le \Phi$ / $0 < (C_P - C_G) \le \Phi$: $d > 0$ YES / NO | $d \le \Phi$ YES / NO | IS THE ROCKET STABLE? YES / NO |
|---|---|---|---|---|---|---|---|---|
| 1 | 1 | 10 | 8 | 2 = 10 - 8 | YES | YES | NO | |
| 2 | | 10 | 8 1/4 | | | | | |
| 3 | | 10 | 8 1/2 | | | | | |
| 4 | STEP 4 | 10 | 8 3/4 | STEP 4 | | | STEP 4 | |
| 5 | | 10 | 9 | | | | | |
| 6 | | 10 | 9 1/4 | | | | | |
| 7 | 1 | 10 | 9 1/2 | | | | | |
| 8 | 1 | 10 | 9 3/4 | | | | | |
| 9 | 1 | 10 | 10 | | | | | |
| 10 | 1 | 10 | 10 1/4 | | | | | |
| 11 | 1 | 10 | 10 1/2 | | | | | |
| 12 | 1 | 10 | 10 3/4 | | | | | |
| 13 | 1 | 10 | 11 | | | | | |

**Figure 15: Center of Gravity and the Center of Pressure diameter, $\Phi$, check**

# INSTRUCTIONS FOR FILLING OUT OUR ROCKET STABILITY TABLE:

## STEP 5 of 5:

Now we fill in the last column.

*If the answer to step 2 was YES, then we can put a YES in this column.*

***THIS ROCKET IS STABLE***

*If the answer to step 2 was NO, then we put a NO in this column.*

***THIS ROCKET IS UNSTABLE***

**Note**:

**If the center of gravity, $C_G$, is NOT less than or equal to the center of pressure, $C_P$, the rocket is *unstable!***

In Step 2 of 5 we noted that $C_G < C_P$. This can only be true if d > 0. So if you check step 2 with a YES, step 3 is also a YES. So, we could skip all of the other steps and put a YES in Step 5. Remember, Step 4 is used as a rule of thumb and can give us better designs if we follow it. Later, we'll discuss fuel burn in ***Section 4.6*** and its effects on the movement of $C_G$ during flight.

Go ahead and continue down each row following the **5 steps** we used to fill in the first row. You can use the table in ***Figure 17***.

| ROCKET # | Diameter of rocket inches $\phi$ Given to us | Distance from tip to $C_P$ inches $C_P$ Given to us | Distance from tip to $C_G$ inches $C_G$ Given to us | Distance from $C_P$ to $C_G$ inches $d = C_P - C_G$ We calculate | IS $C_G < C_P$ YES / NO | IS $0 < d \leq \Phi$, $0 < (C_P - C_G) \leq \Phi$: $d > 0$ YES / NO | $d \leq \Phi$ YES / NO | IS THE ROCKET STABLE? YES / NO |
|---|---|---|---|---|---|---|---|---|
| 1 | 1 | 10 | 8 | 2 = 10 - 8 | YES | YES | NO | YES |
| 2 | 1 | 10 | 8 1/4 | | | | | |
| 3 | 1 | 10 | 8 1/2 | | | | | |
| 4 | 1 | 10 | 8 3/4 | | | | | |
| 5 | 1 | 10 | 9 | | | | | |
| 6 | 1 | 10 | 9 1/4 | | | | | |
| 7 | 1 | 10 | 9 1/2 | | | | | |
| 8 | 1 | 10 | 9 3/4 | | | | | |
| 9 | 1 | 10 | 10 | | | | | |
| 10 | 1 | 10 | 10 1/4 | | | | | |
| 11 | 1 | 10 | 10 1/2 | | | | | |
| 12 | 1 | 10 | 10 3/4 | | | | | |
| 13 | 1 | 10 | 11 | | | | | |

(Arrows labelled "STEP 5" point up to each of the last four columns.)

**Figure 16: Check for Rocket Stability Based on Rules**

| ROCKET # | Diameter of rocket inches $\phi$ Given to us | Distance from tip to $C_P$ inches $C_P$ Given to us | Distance from tip to $C_G$ inches $C_G$ Given to us | Distance from $C_P$ to $C_G$ inches $d = C_P - C_G$ We calculate | IS $C_G < C_P$ YES / NO | IS $0 < d \leq \Phi$ $0 < (C_P - C_G) \leq \Phi$ $d > 0$ YES / NO | $d \leq \Phi$ YES / NO | IS THE ROCKET STABLE? YES / NO |
|---|---|---|---|---|---|---|---|---|
| 1 | 1 | 10 | 8 | 2 = 10 - 8 | YES | YES | NO | YES |
| 2 | 1 | 10 | 8 1/4 | | | | | |
| 3 | 1 | 10 | 8 1/2 | | | | | |
| 4 | 1 | 10 | 8 3/4 | | | | | |
| 5 | 1 | 10 | 9 | | | | | |
| 6 | 1 | 10 | 9 1/4 | | | | | |
| 7 | 1 | 10 | 9 1/2 | | | | | |
| 8 | 1 | 10 | 9 3/4 | | | | | |
| 9 | 1 | 10 | 10 | | | | | |
| 10 | 1 | 10 | 10 1/4 | | | | | |
| 11 | 1 | 10 | 10 1/2 | | | | | |
| 12 | 1 | 10 | 10 3/4 | | | | | |
| 13 | 1 | 10 | 11 | | | | | |

**Figure 17: Blank Stability Table For Student to Complete**

***Figure 23*** illustrates the completed table showing the results for each of the rows.

Following the first table, you'll see another illustration that summarizes the results from our calculations. ***See Figure 24***.

The ***top section*** of the table predicts a **good stable design**, but the distance from the center of gravity to the center of pressure is larger than the diameter of the rocket.

*We can fly this rocket, but the designers have asked us to keep the distance between the center of gravity and the center of pressure* ***less than or equal*** *to the diameter, but greater than zero.* The distance is greater than zero, but it is *NOT* ***less than or equal*** *to the diameter.*

So, even though the rocket is stable, we note this condition on our table in ***Figure 24***.

The ***middle section*** of the table shows us our ***best stable design***. In these rows, both of the good conditions exist. The rocket is stable, *AND* the distance between the center of pressure and the center of gravity is less than the diameter.

The ***bottom section*** of the table, well; you guessed it. In these rows we get the "***Oh Yikes***" condition.

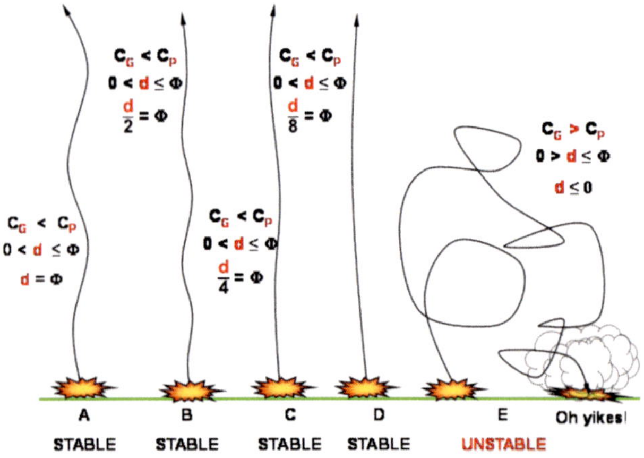


**Figure 18: Results from our table**

## FUEL BURN AND THE EFFECTS ON $C_G$

At this point, it is good to think about one more thing that can cause a rocket to be less efficient.

When designing a rocket, you have to pay attention to the changes in the center of gravity during the ascent stage as fuel is consumed. The rule for keeping the distance between the center of gravity and the center of pressure less than or equal to the diameter, but greater than zero will normally take care of this change in mass and thus center of gravity changes due to fuel consumption.

During rocket ascent rocket fuel is being consumed. This causes the $C_G$ to move towards the tip of the rocket. At the same time, there is an increase in the distance between $C_G$ and $C_P$.

As the distance between the center of gravity and the center of pressure increase, the rocket has a tendency to not fly as straight as it did just after launch

If you make two tables you can check to see this behavior.

The first table is created from a fulcrum test with a rocket that has a motor that is fully fueled (full of propellant).

A second table is created from a fulcrum test with a rocket that has been fitted with an empty motor (no propellant).

Look at ***Figures 21 and 22*** and you will see this how this distance between the center of gravity and the center of pressure change in flight due to the loss of fuel. ***Figure 18*** shows how the change in mass during a flight can change how well the rocket flies. As fuel is burned, the distance between $C_G$ and $C_P$ will increase causing the rocket to oscillate slightly more.

**DO NOT ATTEMPT TO FLY UNSTABLE ROCKETS!**

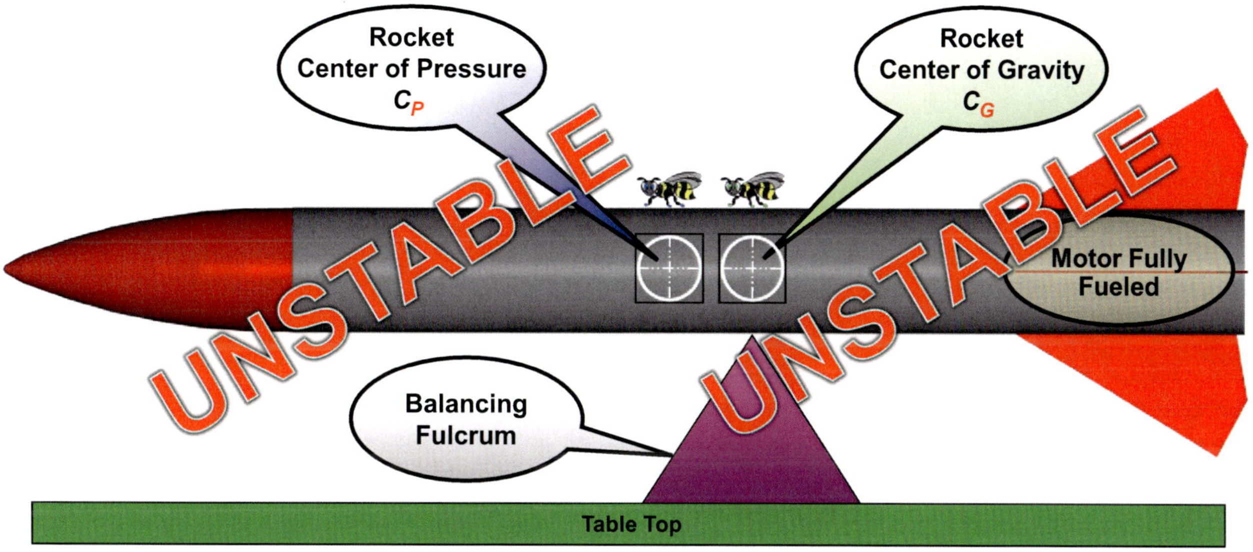


**Figure 19: Rocket Center of Gravity BEFORE Adding Payload**

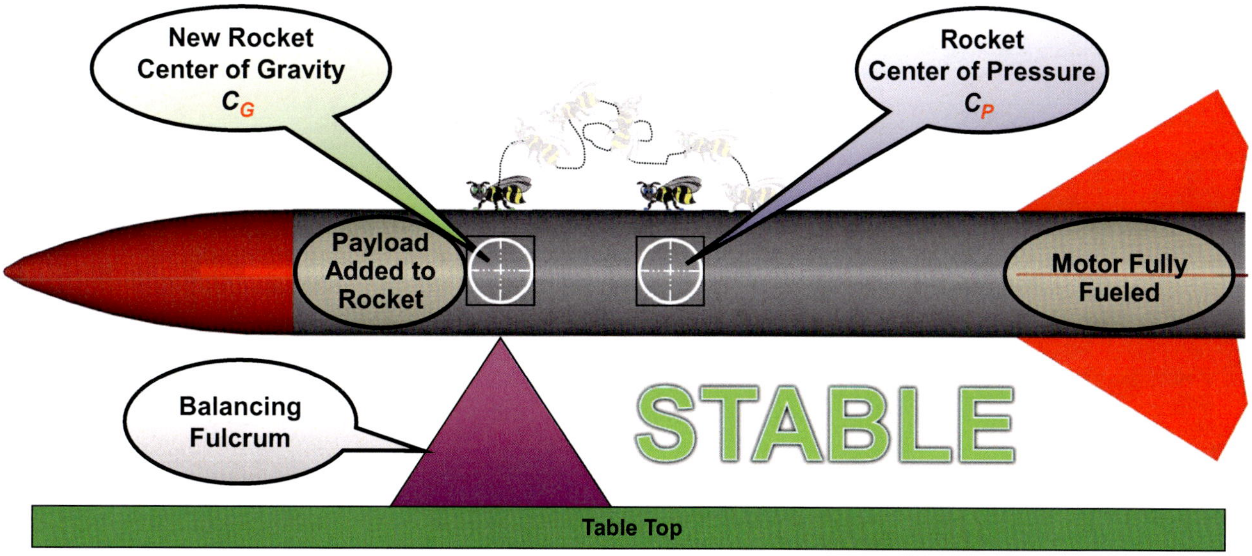


**Figure 20: Rocket Center of Gravity AFTER Adding Payload**

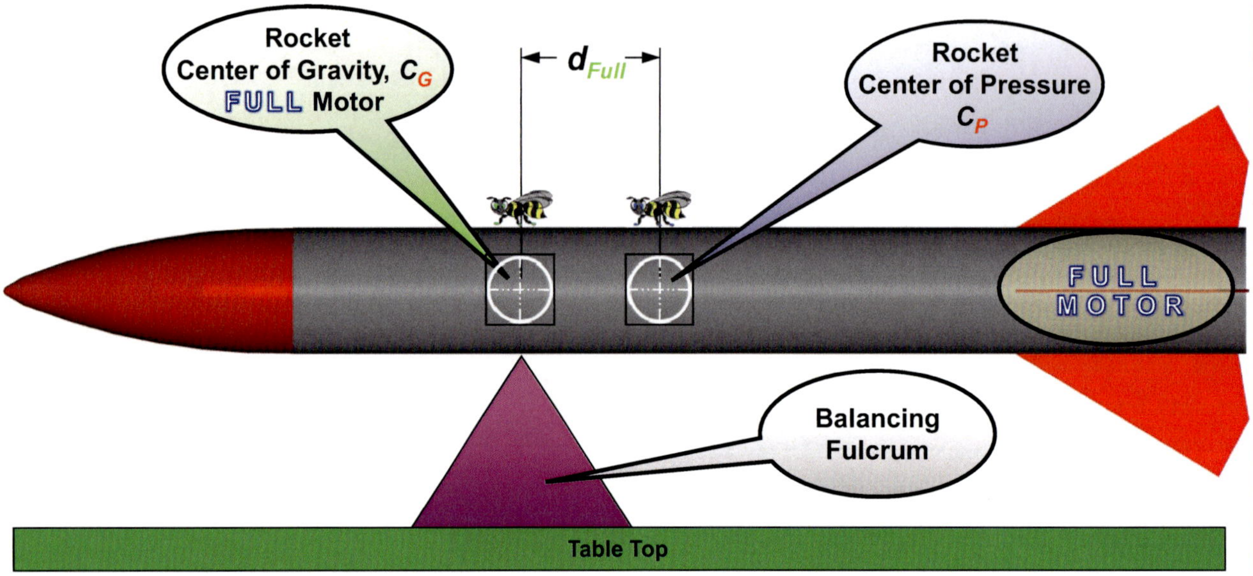


**Figure 21: Rocket Center of Gravity with FULL Propellant**

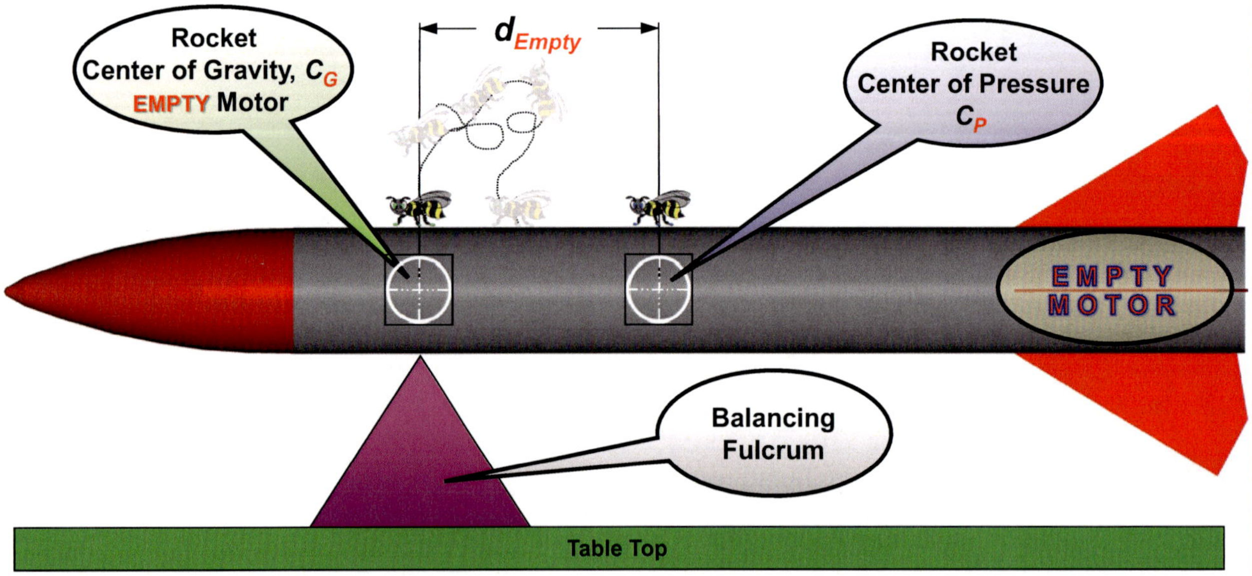


**Figure 22: New Center of Gravity with EMPTY (Spent) Propellant**

| ROCKET # | Diameter of rocket inches $\phi$ Given to us | Distance from tip to $C_P$ inches $C_P$ Given to us | Distance from tip to $C_G$ inches $C_G$ Given to us | Distance from $C_P$ to $C_G$ inches $d = C_P - C_G$ We calculate | IS $C_G < C_P$ YES / NO | IS $0 < d \leq \Phi$ $0 < (C_P - C_G) \leq \Phi$ $d > 0$ YES / NO | $d \leq \Phi$ YES / NO | IS THE ROCKET STABLE? YES / NO |
|---|---|---|---|---|---|---|---|---|
| 1 | 1 | 10 | 8 | 2 | YES | YES | NO | YES |
| 2 | 1 | 10 | 8 1/4 | 1 3/4 | YES | YES | NO | YES |
| 3 | 1 | 10 | 8 1/2 | 1 1/2 | YES | YES | NO | YES |
| 4 | 1 | 10 | 8 3/4 | 1 1/4 | YES | YES | NO | YES |
| 5 | 1 | 10 | 9 | 1 | YES | YES | YES | YES |
| 6 | 1 | 10 | 9 1/4 | 3/4 | YES | YES | YES | YES |
| 7 | 1 | 10 | 9 1/2 | 1/2 | YES | YES | YES | YES |
| 8 | 1 | 10 | 9 3/4 | 1/4 | YES | YES | YES | YES |
| 9 | 1 | 10 | 10 | 0 | NO | NO | YES | ????? |
| 10 | 1 | 10 | 10 1/4 | - 1/4 | NO | NO | YES | NO |
| 11 | 1 | 10 | 10 1/2 | - 1/2 | NO | NO | YES | NO |
| 12 | 1 | 10 | 10 3/4 | - 3/4 | NO | NO | YES | NO |
| 13 | 1 | 10 | 11 | -1 | NO | NO | YES | NO |

**Figure 23: Table to check stability of a rocket with all columns completed**

| ROCKET # | Diameter of rocket inches $\phi$ Given to us | Distance from tip to $C_P$ inches $C_P$ Given to us | Distance from tip to $C_G$ inches $C_G$ Given to us | Distance from $C_P$ to $C_G$ inches $d = C_P - C_G$ We calculate | IS $C_G < C_P$ YES / NO | IS $0 < d \leq \Phi$ $0 < (C_P - C_G) \leq \Phi$ $d > 0$ YES / NO | $d \leq \Phi$ YES / NO | IS THE ROCKET STABLE? YES / NO |
|---|---|---|---|---|---|---|---|---|
| 1 | 1 | 10 | 8 | 2 | YES | YES | NO | YES |
| 2 | [illegible] | [illegible] | 8 1/4 | [illegible] | [illegible] | [illegible] | [illegible] | [illegible] |
| 3 | 1 | 10 | 8 1/2 | [illegible] | YES | YES | NO | YES |
| 4 | 1 | 10 | 8 3/4 | 1 1/4 | YES | YES | NO | YES |
| 5 | 1 | 10 | 9 | 1 | YES | YES | YES | YES |
| 6 | 1 | 10 | [illegible] | [illegible] | [illegible] | [illegible] | YES | YES |
| 7 | 1 | 10 | 9 1/2 | 1/2 | YES | YES | YES | YES |
| 8 | 1 | 10 | 9 3/4 | 1/4 | YES | YES | YES | YES |
| 9 | 1 | 10 | 10 | 0 | NO | NO | YES | ????? |
| 10 | 1 | 10 | 10 1/4 | - 1/4 | NO | NO | YES | NO |
| 11 | 1 | [illegible] | [illegible] | [illegible] | [illegible] | [illegible] | [illegible] | NO |
| 12 | 1 | 10 | 10 3/4 | - 3/4 | NO | NO | YES | NO |
| 13 | 1 | 10 | 11 | -1 | NO | NO | YES | NO |

*STABLE DESIGN BUT d IS NOT LESS THAN or EQUAL TO THE DIAMETER*

**BEST *STABLE* DESIGN**

***UNSTABLE DESIGN...Oh Yikes!***

**Figure 24: Stability and design check for different payload locations**

**Compare your table with the table in *Figure 23*. If your answers are the same, you're well on your way to becoming a rocket scientist. Now we're ready to create a lab to demonstrate how we can use this table to help us design our own stable rocket.**

### 4.6 Create a lab to verify or deny the work in Steps 1-5

1. Draw a picture of the problem
2. Label the picture
3. Write a mathematical expression that defines the picture
4. Solve for the unknowns
5. Calculate the results (by hand, by calculator, or by computer application software)
6. **Create a lab to verify or deny the work in 1-5**
7. Test the results of 1-5 in the lab

In a real rocket, one of the first things to consider is "what will the payload be?" In some rockets, it may just be a parachute to bring the rocket back to earth safely. In others, the payload could be people and equipment needed to perform experiments in space. Once the payload is determined, the rest of the rocket is designed around it. Once we know the size and weight of the payload, a motor is selected that will provide enough thrust to safely get the rocket to its destination. The payload and rocket motor together with all the necessary body structures such as the nosecone, fuselage (payload area), motor housing and fins make up the final design.

In many cases, after a rocket has already been designed, more payload weight may have to be added. Based on the location of the additional weight relative to the nosecone tip, the $C_G$ dimension may change. Usually, we just check to see if the $C_P$ is still greater than the $C_G$.

If it is, we won't have to make changes to the rest of the rocket fuselage, fins or nosecone. If the $C_G$ is greater than the $C_P$, the rocket can be made longer, or perhaps the fins can be made larger. This adds even more weight to the rocket, requiring more fuel and thus again making the rocket even heavier. The more fuel we use will make the rocket cost more per launch, so we have to pay very close attention to the placement of our payload.

One more thing we should discuss. When the rocket starts to fly, fuel is used. This really helps us because not only is the rocket getting lighter, but the center of gravity, $C_G$, is actually moving closer to the nosecone tip, which guaranties stability.

The lab we are about to do will demonstrate some of these principles. We will be using a rocket that has already been made and only has a parachute as its payload.

The engineers have asked us if they can add a video camera, transmitter and transmitting antenna along with batteries to power the electronics in the rocket. The engineers have told us that they have enough fuel to get the rocket up, but with the added parts, it just won't go as high. Even though they have enough fuel, they're not sure the added payload will affect the stability or size of the rocket .

What we need to do next is determine where we can place the added payload while keeping the rocket stable. Secondly, we need to determine if the rocket can stay the same size with the added payload

By following the instructions below, we should be able to determine where to place the payload and still have a stable rocket.

Read the following procedures very carefully and see if you can determine where we can place our payload. Our cutout model is one half the size of the real rocket, so the center of pressure that we measure will have to be multiplied by 2 to get our actual center of pressure for the real rocket.

**Below is the procedure for LAB 3:**

We are going to do something very similar to the steps in LABS 1 & 2. We are going to cut out the pieces of the rocket from paper and tape them together.

When we're done though, we are going to glue the cutout pieces onto some cardboard. Finally we are going to cut the cardboard from around the paper. We will use this cardboard cutout in a clever way to *ESTIMATE* the $C_P$.

***Note: The cutout drawings provided in the appendix are for the LOC IV. A link to this rocket kit has been provided in the Section 7.*** *This drawing represents the "shadow" of the LOC IV rocket as if it were projected onto a wall. This shadow when traced onto a piece of paper will form our cutout.* ***It's up to you to decide if your rocket is safe or not!***

***If you don't have a LOC IV rocket, you can follow these steps for any rocket you have.***

1. **Cut out the three sections of the rocket from the drawings provided or your own rocket using the projection method mentioned above**
2. **Tape all the pieces together like you did in LAB1.**
3. **Glue your finished paper model onto a piece of cardboard.**
4. **Cut the cardboard around the rocket outline until you have your finished cardboard rocket.**

This cardboard model can now be used to get a good estimate of the rocket's actual center of pressure, $C_P$. The drawings provided for the LOC IV are ½ the actual size of the rocket. It's what we call a scaled version. This one is a **one half scale** version of the real rocket.

See if you can balance the cardboard cutout on the edge of a ruler. It may help to have one person holding the ruler, while another person balances the cardboard cutout. *When the rocket is balanced, mark the location on the cutout with a sharp pencil using the ruler as a straight edge*.

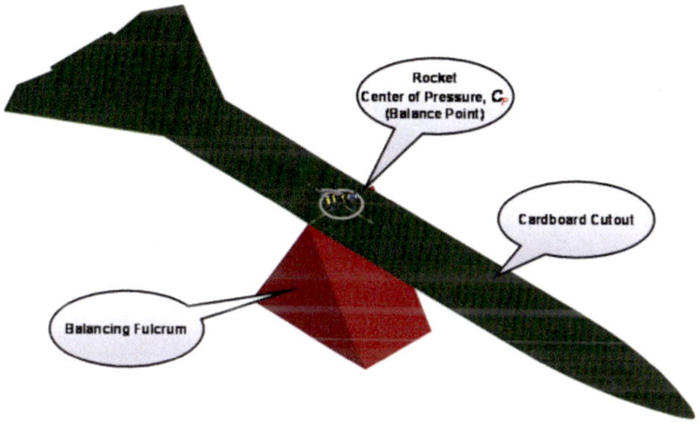


**Figure 25: Locating the Center of Pressure, $C_P$**

Do this a few more times just to be certain your mark is in the right location.

**If you're using the LOC IV cutout, remember it is one half the size of the rocket. The measurement you get will have to be multiplied by two to get the actual $C_P$ for the LOC IV.**

If you're using your own rocket, measure the distance from the mark you just made to the tip of the nosecone. Put your measurement in the space provided below.

**How far is the $C_P$ from the nosecone tip? Don't forget your units!**

______________________________

______________________________

**Congratulations. You just found the $C_P$!**

Okay. We're halfway there. We know our center of pressure, $C_P$, and now we are going to find the center of gravity, $C_G$. This is easy.

See if you can balance your rocket by using the side of a ruler. Be careful. One student can hold the nosecone side of the rocket, while another student holds the motor end. A third and fourth student can hold each side of the ruler on its edge.

Make sure you have a fully fueled motor in your rocket before taking this measurement.

Now, balance the rocket on the ruler and place a piece of tape on the rocket to mark where the center of gravity is located.

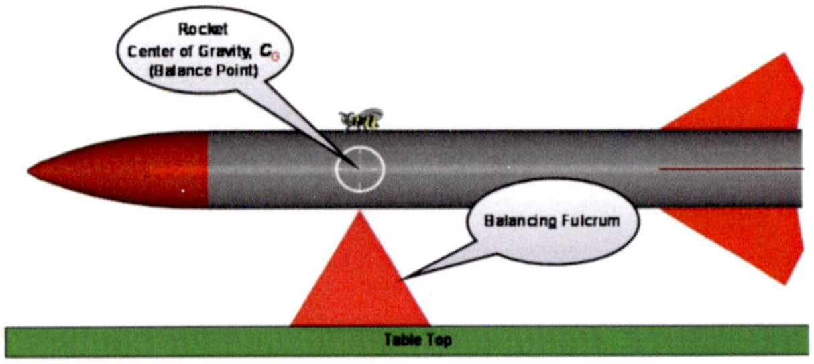


**Figure 26: Finding the center of gravity, $C_G$**

Do this several times to make sure you've marked the rocket properly.

**How far is the $C_G$ from the nosecone? Don't forget your units!**

**Congratulations, you've just found the $C_G$!**

**What will happen to the $C_G$ after the fuel is consumed in flight?**

The rocket you just balanced had a motor that was full of fuel. **Remember, as the fuel burns the rocket gets lighter and the $C_G$ changes.** Replace the "full of fuel" motor from the motor section of the rocket with the "spent" motor that does not have fuel. Check the $C_G$ again just as you did the first time. (This takes four students.) Place a piece of tape to mark where the $C_G$ is for the "spent" motor.

**What happened to the $C_G$ after you measured the rocket with a "spent" empty motor? Did it do the same thing as the illustrations in *Figures 21 & 22*?**

**Start with a fully fueled rocket. From the payload section of your rocket, add a payload weight. You can use tape to hold the weight in position. Recheck the $C_G$. Continue down the rocket this way until you find out where the weight would cause your rocket to be unstable. Write down the weight of the payload and the location on the rocket that caused your rocket to be unstable.**

## 4.7 Test the results of 1-5 in the lab.

1. Draw a picture of the problem
2. Label the picture
3. Write a mathematical expression that defines the picture
4. Solve for the unknowns
5. Calculate the results (by hand, by calculator, or by computer application software)
6. Create a lab to verify or deny the work in 1-5
7. **Test the results of 1-5 in the lab**

How are we going to test our lab? One sure way is to take the rocket out and fly it.

Sometime in the spring, ask your teacher to fly this rocket. Your teacher will let you know when and where. Remember that we want to fly our rocket on a nice warm, calm sunny day.

**SAFTEY TIPS**:

**Rockets can never be flown if wind speeds are greater than or equal to 20 miles an hour.** So be patient, wait for good weather, and test your skills as rocket designers with a test launch.

The *NAR, National Association of Rocketry* as well as the the *FAA, Federal Aviation Administration* are wonderful organizations that provide information on safety, certification and regulation. Links to both of these organizations are provided in ***Section 7***.

One of the things you'll find most useful before during, and after a launch are Checklists.

***Figures 28-32*** illustrate several checklists used for "Josie", the rocket on the cover of this book. The checklists for Josie can be used as a guide for your own checklists.

***Figure 33*** is a blank checklist that may help you get started. It can be used to create your own checklist. Snacks and drinks are great checklist items!

**HAVE FUN!**

Taking video of your flight is a great way to review the launch. With your video, you will be able to review the performance of your rocket after the flight. If you're lucky, you should get some nice pictures too!

With an actual flight, you will be able to confirm your stability checks.

After you verify your launch, ask your teacher to fill in the information on your class Rocket Stability Certificate.

Fill in the NAME of the rocket. The name of the rocket on the cover of this book is "JOSIE".

Fill in your $C_G$ and the $C_P$ measurements on your certificate.

Fill in the date of your launch, your school name and your teacher's name.

***Don't forget to sign your name on the bottom of the certificate.***

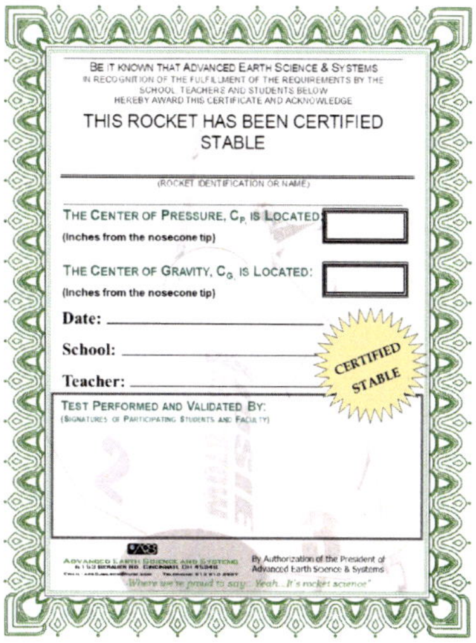


**Figure 27: Rocket Stability Certificate**

Once you've filled in all the information on the certificate provided by your teacher, take a class photograph and display the photograph alongside both your rocket and your signed certificate.

| EQUIPMENT LIST (1 of3 ) | | EQUIPMENT LIST (2 of 3) | | EQUIPMENT LIST (3 of 3) | |
|---|---|---|---|---|---|
| Comments | Comments | Comments | Comments | Comments | Comments |
| FIRE EXTINGUISHER | Charged (On Pre-FLT List) | MEASUREMENT | LAPTOP | Field COMM radios (MIC HEADSETS) | batteries charged |
| FIRST AID KIT | Full (On Pre FLT List) | | Barometer, temp, humidity gage | | turned off |
| DISPOSAL CONT. | FIREPROOF | | Anemometer | Field VID CAMERA | battery charged |
| STP motor cleaners | ☐ | | Wind Vane | | video media |
| paper towels | ☐ | | binoculars | Field STILL camera | card empty |
| motor disposal bags | ☐ | | inclinometer tool | | batteries charged |
| LAB COAT | ☐ | | VOM | Field GPS | batteries charged |
| LAB BOOK | ☐ | | stopwatch | CELL phone | batteries charged |
| ROCKET | Parasail and E-Chute | | calculator | Equipment table | ☐ |
| MOTOR | IGNITERS | Toolbox (MIN) | scissors, Xacto knife | A/V | Ground Rec Antenna |
| LAB BOOK | ☐ | | 100' Tape measure | | Video Monitor |
| Launch Pad | tower | | tape, electrical, masking | | Tape REC/PLAY |
| | Hor. siren/beacon CNTDWN rail | | epoxy, CA glue | | Tripod |
| | Vertical Launch rail | | Allen wrenches (door latch) | | VIDEO AMP |
| | flags | | hemostats | | Projector |
| | beacon and CNT DWN display | | Small Philips screwdriver | | Projector Screen |
| | parasail pack fixture | | Electric screw driver with Philips | | PV amp |
| MEGAPHONE(SIREN) | Batteries Charged | | mineral oil (Door latches) | | two(2) RCA to 1/4 audio cables |
| platform base | platform hardware | | petroleum jelly ("O" Rings) | | Speakers |
| ROCKET VID TRANS | Charged | | steel wool (in zip lock) | PLC Controller | 6 AMP portable inverter |
| ROCKET STILL CAM | card empty | | safety key | | control conduit |
| | batteries charged | | bunji cords | | 12VDC Ignition Battery |
| ROCKET Servo BAT | Charged | | sonic detector and battery | | motion sensor |
| | turned off | | canopy line tool | | safety key |
| T6XA 50MHz w/Flag | Charged | | power supply cable | | 12VDC to alligator launch cable |
| | turned off | | igniter cable | | 12VDC inverter battery |
| T6XA 50MHz w/Flag | Charged | | insulation & 4" template | | PLC AC power line |
| | turned off | | ☐ | | PLC AC extension cord |

**Figure 28: Equipment Checklist**

| Date: Time: Temp (F): | Pressure (Hg): Humidity(%): |
|---|---|
| Rocket: RS09870W-JOSE | Winds(mph) & From Dir: |
| Pre Flight Checklist (General Rocket Condition) | |
| Comments | Comments |
| ☐ lines checked (tangles, unattached or broken split rings) | ☐ barometric pressure, temp, humidity, winds noted |
| ☐ aileron cables (worn, broken) | ☐ Remove and secure SAFETY key |
| ☐ motor packed(O-rings, tape on forward end) | ☐ sequencer power ON |
| ☐ motor secured | ☐ 900' servo radio check |
| ☐ launch rail check | ☐ |
| ☐ rail guide check | ☐ |
| ☐ nose cone secure | ☐ |
| ☐ fin check | ☐ |
| ☐ canopy module secure | ☐ |
| ☐ e-chute module secure | ☐ |
| ☐ motor module secure | ☐ |
| ☐ fuselage insulation | ☐ |
| ☐ FIRE EXTINGUISHER ACCESS AND CHARGED | ☐ |
| ☐ FIRST AID KIT ACCESS | ☐ |

**Figure 29: Pre Flight Checklist General Rocket Condition**

| Date: | Time: | Temp (F): | Pressure (Hg): | Humidity(%): |
|---|---|---|---|---|
| Rocket: RS09870W-JOSIE | | | Winds(mph) & From Dir: | |
| Pre Flight Checklist (POWER UP) | | | | |

| Comments | Comments |
|---|---|
| ☐ *CHARGE ROCKET ON RAIL* | ☐ GPS check |
| ☐ Sound check (headsets and mikes) | ☐ Range cameras ready |
| ☐ Video camera on (433Mhz) | ☐ check remote launch sequence servo (50Mhz) |
| ☐ Video transmitter on (433Mhz) | ☐ LAUNCH SAFETY BEACON/CNTDWN/MOTION/SIREN |
| ☐ Video monitor check | ☐ Call Tower-859 372 64 52 |
| ☐ DVD monitor and record check | ☐ REMOVE CHARGING JACK FROM ROCKET |
| ☐ Check Parasail, E-chute and Stabilizer Deploy Servos | ☐ Power ON Servos, Video camera and Transmitter |
| ☐ Check Parasail Flaperons and Stabilizer operation | ☐ Power ON 50 and 72 Mhz servo transmitters |
| ☐ Camera turned on (slide cover) | ☐ Secure e-bay |
| ☐ Check STILL and VID cradle rotation | ☐ Secure canopy bay and stabilizer |
| ☐ Check STILL shutter operation | ☐ Set canopy brake 72Mhz transmitter |
| ☐ Power OFF Servos, Video camera and Transmitter | ☐ ANNOUNCE FIRST AID LOC |
| ☐ Power OFF 50 and 72 Mhz servo transmitters | ☐ ANNOUNCE FIRE EXTINGUISHER LOC |
| ☐ Pack e-chute | ☐ AREA CLEAR |
| ☐ Pack canopy | ☐ Connect Igniter |

**Figure 30: Pre Flight Checklist Power UP**

| Date: | Time: | Temp (F): | Pressure (Hg): | Humidity(%): |
|---|---|---|---|---|
| Rocket: RS09870W-JOSIE | | | Winds(mph) & From Dir: | |
| FLIGHT | | | | |

| Comments | Comments |
|---|---|
| ☐ INSERT SAFETY KEY | ☐ Camera Rotation |
| ☐ SET THUMBWHEEL TO 9s | ☐ Fly Towards Pad 10s 20ft/s +33s -520ft |
| ☐ Remote Sequence Start Ready (50Mhz) | ☐ Deploy E-chute @ 100' @ 5ft/s +33s -520ft (OPT) |
| ☐ Flight Ready (72Mhz) | ☐ Chute Full Deployment @ 50ft +37s-570 ft 5ft/s |
| ☐ Digital Camera Shutter Activate and Lock | ☐ E-chute 8s @ 5ft/s -610ft |
| ☐ Remote Sequence Start (50Mhz) T-9s | ☐ Touchdown @ +40-45s |
| ☐ Liftoff | ☐ |
| ☐ Apogee (+600-700 ft @ 6.3 s) | ☐ REMOVE SAFETY KEY |
| ☐ Canopy Deployment (apogee -50ft @ +10 s) | ☐ |
| ☐ Canopy Full Deployment @ -180ft +13* | ☐ |
| ☐ Remove Brake (apogee -180ft @ +13 s) -180ft | ☐ |
| ☐ Left Turn Descent @ 20 ft/s for 5s +18s - 280' | ☐ |
| ☐ Right Turn Descent @ 20 ft/s for 5s +23s -380ft | ☐ |
| ☐ | ☐ |
| ☐ | ☐ * If canopy malfunction deploy e-chute @ T+13 |
| ☐ | ☐ |

**Figure 31: Flight Checklist**

| | |
|---|---|
| **Date:** **Time:** **Temp (F):** | **Pressure (Hg):** **Humidity(%):** |
| **Rocket: RS09870W-JOSIE** | **Winds(mph) & From Dir:** |
| **Post Flight checklist** | |
| **Comments** | **Comments** |
| ☐ **Safety Key Removed** | ☐ **Motor Housing System (damage, secure)** |
| ☐ **Remove Motor Assembly** | ☐ **lines checked (tangles, broken split rings)** |
| ☐ **Video Camera off (433Mhz)** | ☐ **aileron cables (worn, broken)** |
| ☐ **Camera Turned Off (slide cover)** | ☐ **port door condition, damage** |
| ☐ **Launch Sequence Servo Power Off (50Mhz)** | ☐ **starboard door condition , damage** |
| ☐ **Launch Sequence Transmitter Off (50Mhz)** | ☐ **fin damage** |
| ☐ **Flight Receiver Power Off (72Mhz)** | ☐ **launch lug damage** |
| ☐ **Flight Transmitter Power off (50Mhz)** | ☐ **nose cone secure** |
| ☐ **Parasail Damage Inspection** | ☐ **deployment module secure** |
| ☐ **E-Chute Damage Inspection** | ☐ **fuselage module secure** |
| ☐ | ☐ **Motor Housing Spent Fuel Disposal and Prep** |

**Figure 32: Post Flight Checklist**

Date: Time: Temp (F): Pressure (Hg): Humidity(%):

Rocket: Winds(mph) & From Dir:

Checklist

| Comments | Comments |
|---|---|
| ☐ | ☐ |
| ☐ | ☐ |
| ☐ | ☐ |
| ☐ | ☐ |
| ☐ | ☐ |
| ☐ | ☐ |
| ☐ | ☐ |
| ☐ | ☐ |
| ☐ | ☐ |
| ☐ | ☐ |
| ☐ | ☐ |
| ☐ | ☐ |
| ☐ | ☐ |
| ☐ | ☐ |
| ☐ | ☐ |
| ☐ | ☐ |
| ☐ | ☐ |
| ☐ | ☐ |
| ☐ | ☐ |
| ☐ | ☐ |

**Figure 33: BLANK Checklist**

## 5 Results

Although we will not be able to see the results of this lab until we launch our rocket, we can make some predictions about our flight based on what we have measured. Since the $C_G$ leads the $C_P$, we should be able to predict a safe and beautiful flight.

The mathematics we perform to predict the behavior of a rocket give us the **theoretical results**. The ***empirical results*** are what we will actually observe from the launch.

**Theoretical Results: Predicted**

**Empirical Results: Observed**

## 6 Conclusions (What did we learn?)

- In this lab, we created a mathematical expression to check the stability of a rocket
- Remember that we created that mathematical expression from a picture.
- We learned that we could check the center of gravity and the center of pressure by balancing.
- We learned that "a mathematical expression that defines a picture" can be an **inequality ( <, >, ≥, ≤ )**,or an **equality (=)**.

Can you see why, once we create a mathematical expression, it's very convenient to show it in a table?

Now as in every lab, go back to the beginning of this lab and read the abstract again. Does it make more sense now?

**A SIDE NOTE:**

In this LAB, we used the "***Shadow***" concept to find our $C_P$. You can also ***solve*** for the $C_P$ using only a drawing and mathematics.

You can use both when designing rockets and will find both equally effective. The results for either method will give you an accurate measurement for the $C_P$.

The $C_G$ is also solved with two methods, the "***Balance***" method and a ***mathematical*** method.

Again, both will give you very close results when designing rockets.

## 7 References and additional reading

**7.1 The Handbook of Model Rocketry by G. Harry Stine**

**7.2 The Parachute Manual by Dan Poynter**

**7.3 National Association of Rocketry, www.nar.org**

**7.4 Email: nar-hq@nar.org**

**7.5 http://www.nasa.gov**

**7.6 http://www.Spacex.com**

**7.7 https://www.faa.gov/**

**7.8 https://www.siriusrocketry.biz/ishop/loc-loc-iv-1804.html**

# 8 INDEX

# LAB 4: MOTORS AND NEWTON'S LAWS

# LAB 4: MOTORS AND NEWTON'S LAW

4th - 6th Grade Learning Standards Applied to High Powered Rocket Flight and Descent

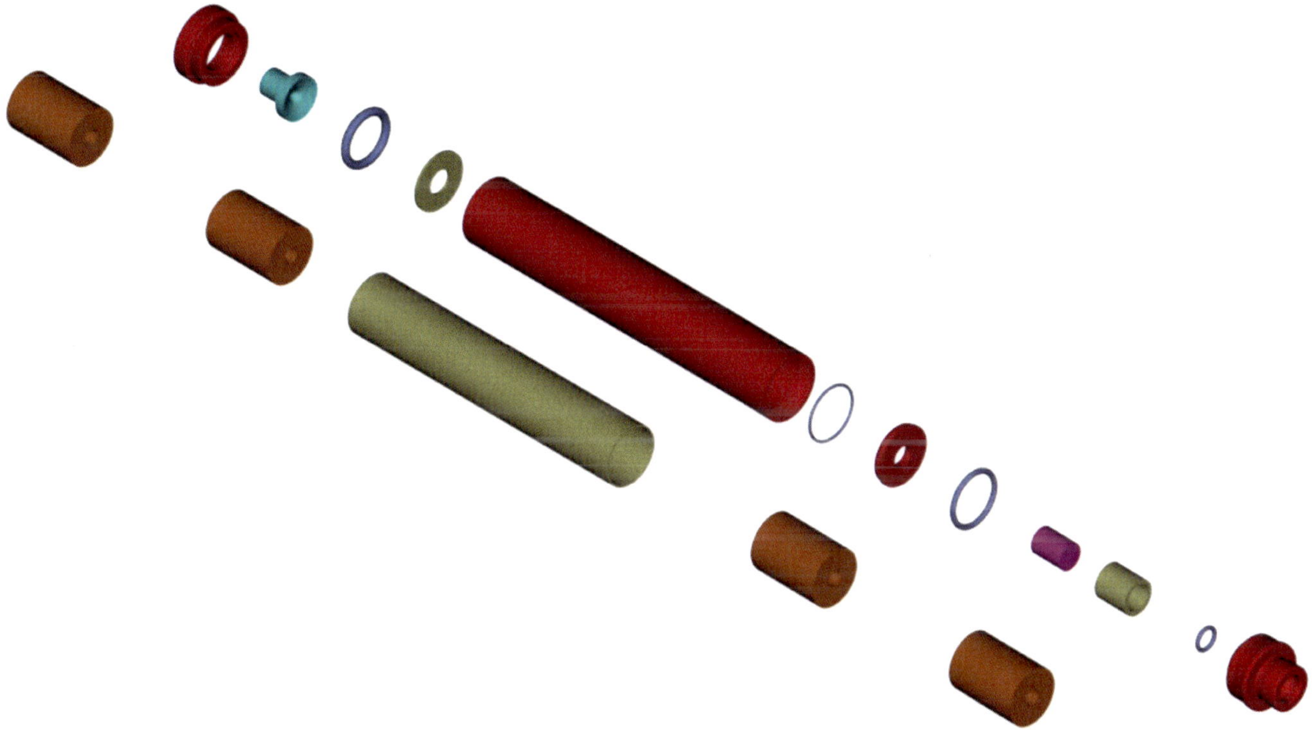

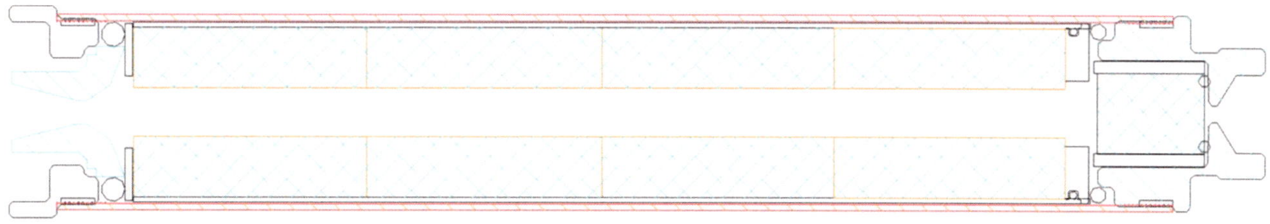

## Table of Contents

## List of Tables and Figures

## 1 Abstract (What are we going to learn?)

Someday soon, a teacher is going to introduce you to some very important mathematical laws that deal with the motion of objects. These laws, called *Newton's Laws,* are very powerful tools for predicting bodies in motion, or at rest.

Sir Isaac Newton was a very curious person. His curiosity and his keen ability to observe events led him to some very remarkable discoveries. These discoveries are what we now refer to as Newton's laws. These laws allow us to predict the behavior of moving objects and objects at rest. Newton's first law could be described in a sentence form like this.

***"A baseball in a mitt remains at rest in a mitt until somebody removes it, or a baseball thrown will remain in motion until it's stopped by the ground or somebody's mitt… dude."***

Sir Isaac Newton (1642-1727) said it like this:

***"A body at rest will remain at rest, or a body in uniform motion in a straight line will maintain that motion, unless an external resultant force acts on it."***

He also said:

***"The time rate of change of momentum of a particle is equal to the resultant force acting on the particle.***

***If two bodies interact, the force of body 1 on body 2 is equal and opposite the force of body 2 on body 1.*** All of rocket science deals with these laws. Using these laws allows us to make jet engines and rocket motors. The applications of these laws allow us to predict what happens when forces interact with each other.

Picture a rocket on a launch pad. It's not going anywhere unless we apply a force to it to make it fly. Newton might say if he were alive today:

***"A rocket on the launch pad will remain on the launch pad unless acted upon by an external force."***

The purpose of this lab is to demonstrate how we can apply these laws to predict how a rocket will fly. ***We will be using a rocket motor as the "external" force.*** With this in mind, this lab will discuss how we can apply an external force that will overcome both the forces of gravity and the forces of air molecules to lift a rocket off the launch pad, and let it soar through the air.

## 2 Glossary of Terms (New words and symbols we need to learn)

**Acceleration**: How fast velocity (speed) changes.

**Aerodynamic**: When an object can speed through the air easily, it is said to be "aerodynamic."

**Apogee:** This is the highest point of a flight, just before descent.

**Altazimuth**: An instrument used to measure vertical angles used for altitude measurements.

**Ascent**: This is the part of flight when the rocket is getting farther away from the earth (flying up).

**Descent**: This is the part of flight when the rocket is coming closer the earth (flying down).

**Cross section**: This is a drawing of something that has been sliced into sections so you can see what is inside. Think of an apple that's been cut in half.

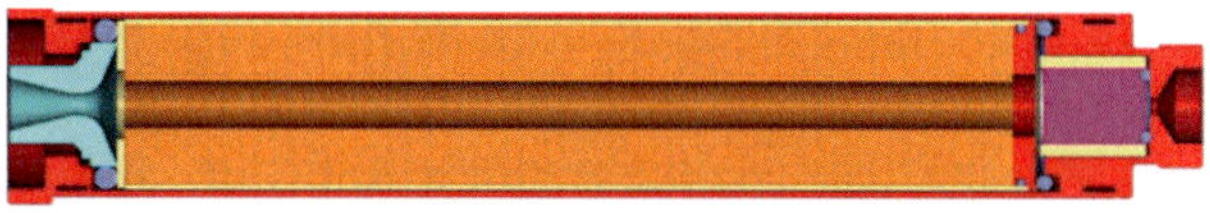

**Figure 1: Rocket motor sliced in half (“*cross section*”)**

$\rho$: This is the Greek symbol rho, used to describe the density of the air. It has units of pounds **mass per cubic inch**, $\frac{lbm}{in^3}$.

$F_g$: This is the symbol used to represent the force of gravity, or weight of an object. It has units of pound force, *lbf*.

$F_m$: This is the symbol used to represent the force of a motor. We call this force “thrust.” It has units of pound force, *lbf*.

$F_t$: This is the symbol used to represent the force of a table acting on an object resting on the table. It has units of *lbf*.

$F_a$: This is the symbol used to represent the force of the air on a rocket in flight. The smaller this force is on our rocket, the better the *aerodynamic* design is. It has units of *lbf*.

**Furlong**: Unit of measurement of Length; 660 feet.

**Fortnight**: Unit of measurement of Time; two weeks.

**Orthogonal**: When an object looks like it's perpendicular, (90 degrees), to another object as you site around it (like a tall tree to the ground).

**Parabolic**: Think of the shape of an egg. A parabola can be defined by a squared function.

**Pressure**: This is a measurement of force per unit of area, pounds per square inch, $lbf/in^2$ or Newtons per square meter, $N/m^2$.

**Speed**: The change in distance per time something has (without regard to the direction)

**Spherical**: Think of a basketball shape. It is also defined by a squared function.

**Velocity**: The change in distance per time including the direction something has.

**Yikes**: A term Hoosiers (people from Indiana) use when something goes terribly wrong. *Oh yikes!*

**Other Symbols used:**

**ϕ** This is a symbol used to represent the **diameter** of our rocket (in inches).

*d* This is a symbol used to represent the **distance between the center of gravity and the center of pressure** of our rocket (in inches).

**=** The equal sign means that two quantities have the same value. This is important when a cake is being cut and you want your = share.

**<** This symbol replaces the phrase **“less than”.** It would be unfortunate if your piece of cake were < your brother's or sister's.

**>** This symbol replaces the phrase **“greater than”.** It would be great if your piece of cake were > your brother's or sister's.

**≤** This symbol replaces the phrase **“less than OR equal”.** This means that a quantity can be less than or equal to another quantity, but cannot be larger. Is it a good thing that your brother's or sister's cake is ≤ to your piece of cake?

**≥** This symbol replaces the phrase **“greater than OR equal”.** This means that a quantity can be greater than or equal to another quantity, but cannot be smaller. Is it a good thing that your brother's or sister's cake is ≥ to your piece of cake?

## 3 Introduction and Purpose (Why are we doing this?)

A ***rocket motor*** must supply a force, or thrust to a rocket that is *greater than the combined forces of gravity and the force of the air molecules bombarding the rocket during flight.*

In this lab, we are going to discuss the rocket motor and how it works. We are also going to discuss aerodynamic shapes. We will explore some of the mathematical expressions that define how high a rocket can fly, given the right motor, and given that the rocket has the right aerodynamic shape.

*Although the forces due to gravity are pretty much the same throughout short, low flights, the forces due to air are not.* *The opposing forces due to air get much bigger as our rocket gains velocity.*

During flight, when our rocket velocities increase, these "pesky" molecules we call "air" really begin to cause the rocket to slow down. That's because the forces against our rocket from these molecules increase as our rocket velocities increase. A force of one body on another body gets bigger if a collision takes place and the change in velocity of a body takes place over a shorter period of time.

Since we are invading the space of the molecules, and they are resisting a change in their motion, we have to work with them. We do this by making rockets, cars, planes or anything else we want to go fast in fluid like air or water "*aerodynamic*."

We do this because the effects of these molecules can be greatly reduced with good *aerodynamic* designs. If we fly high enough, out of the earth's atmosphere, the forces of these pesky molecules go away.

Here, high in the sky, there are very few molecules. Less molecules, lest stuff to fly through. Combine that with the lessening forces due to gravity as we climb farther away from the earth and the loss of weight as the rocket burns through its fuel, cause our rocket velocities to increase dramatically during our ascent.

The forces due to gravity on the rocket actually get smaller as the rocket gets farther from the ground. For short flights, we won't see any noticeable change, but for long flights into space, we will see a dramatic decrease in force due to gravity as the rocket gets farther from the ground.

On the way down though, (we call this "descent"), these pesky little molecules can be very helpful. With the aid of a parachute, these molecules help us to bring our vehicle back down to earth safely by slowing our descent. Air molecules really don't care if you're going up or down, they just like to be in the way and slow anything down that passes through their neighborhood.

Our rockets don't go high enough to see any reduction in the forces due to the molecules, nor are we high enough to see any reductions in the effects of the forces due to gravity. Our rockets during the entire flight and descent are fighting the forces due to gravity and air. We'll leave the really high flying to SpaceX and NASA.

The forces from air can be a little tricky to deal with. But as you will see in this lab, we can out smart these pesky atoms and molecules. The forces of air on a rocket, $\boldsymbol{F_a}$ once mastered will be one of the best tools you can use to improve the performance of a rocket. In the near future, you may see rockets built with nosecones that change shape during flight. For larger rockets, this could be a method to save huge amounts of fuel during ascent. These weight savings can be converted to payload weight or higher faster flights or even reductions in rocket overall sizes.

# 4 Procedure: (How do we do it?)

1. Draw a picture of the problem
2. Label the picture
3. Write a mathematical expression that defines the picture
4. Solve for the unknowns
5. Calculate the results (by hand, by calculator, or by computer application software)
6. Create a lab to verify or deny the work in 1-5
7. Test the results of 1-5 in the lab

**Figure 2: Seven steps to solving a problem**

In this LAB, we are going to draw and label three pictures. The first will be a picture that describes the ROCKET AT REST. The second will be the picture that describes the rocket just as it leaves the launch pad, (LIFTOFF), and the third picture during the ASCENT stage that takes into account the forces of air on the rocket.

## 4.1 Draw a picture of the problem

1. **Draw a picture of the problem**
2. Label the picture
3. Write a mathematical expression that defines the picture
4. Solve for the unknowns
5. Calculate the results (by hand, by calculator, or by computer application software)
6. Create a lab to verify or deny the work in 1-5
7. Test the results of 1-5 in the lab

### 1. PICTURE OF ROCKET AT REST

When a rocket is resting on the launch pad, the only forces we will concern ourselves with is the ***force of gravity*** *(weight of the rocket)* and the ***force of the table***, which holds the rocket. We refer to forces due to gravity as *weight*.

The interesting thing about gravity is that it exists all over this universe and in all other universes.

It's an odd thing to even discuss, because we can't use our senses of smell, sight, hearing or taste to detect it.

For example, a ball released from your hand will fall to the ground. Although we can't use our senses directly to detect gravity, our senses inform us of **the effects of gravity**.

**We see the ball falling but don't actually see the gravity that's making it fall.** It would be like watching a ball being thrown through the air without anyone there to throw it. Weird right?

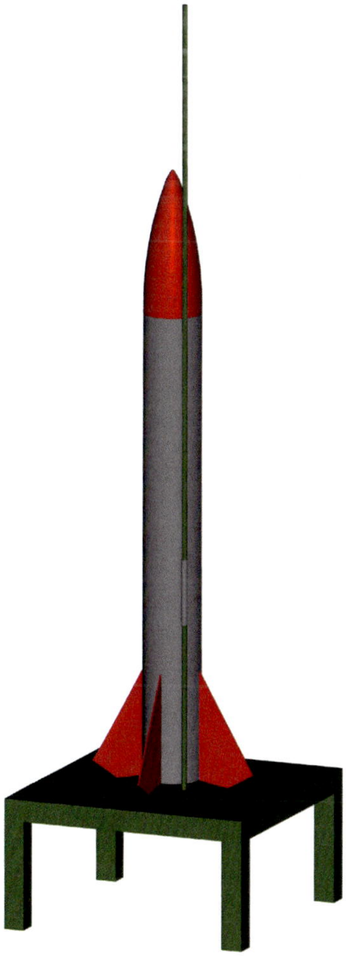

**Figure 3: Rocket On the Launch Pad**

We see and feel only the **effects of gravity.** We use our sense of feel (our tactile sense), and sight. We can do this by

**observing** our surroundings and the effects of gravity that change our surroundings.

Our tactile senses allow us to feel objects and sensations. One way we can "feel" the force of gravity is in our feet. We feel the effects of gravity just by standing or running or even holding something in our hands.

What's even more amazing is that there is a force of attraction between all objects on all other objects. The forces of attraction of one body on another are called gravitational forces. All objects gravitate to one another. This may seem a bit strange, but it's true.

***Look at the person next to you. If you were in space far away from other bodies of mass like planets and stars, you, and the person seated next to you would eventually be drawn together, just by* the gravitational forces of your two bodies.**

Can you feel the forces of the sun, the solar system, the galaxy? Probably not, even though we know they exist.

*Here's a strange thought. If you were in a free falling elevator, even the tactile sense in your feet would disappear. The forces of gravity would still be at work, but you would not be aware of it.*

If we go back and look at the rocket on the launch pad, we see that it is stuck to the table by gravitational forces. The table is holding the rocket and preventing it from falling to the ground. The table resists the weight of the rocket to keep it at rest. If it didn't, well, the table wouldn't be much use to us.

Let's go ahead and draw a picture of a rocket on a launch pad. *See **Figure 4**.* **In this picture, the rocket is not moving relative to the table.** It's waiting on an external force to make it do something.

It's interesting that the table and the rocket **relative** to each other are not in motion. But if you consider the rotation of the earth on its axis, the rotation of the earth in the solar system around the sun, and the motion of our galaxy throughout the universe the launchpad and the rocket are moving through space at over ½ million miles per hour! So are you!

**Here the rocket relative to the launchpad will remain at rest until an external force acts on it.**

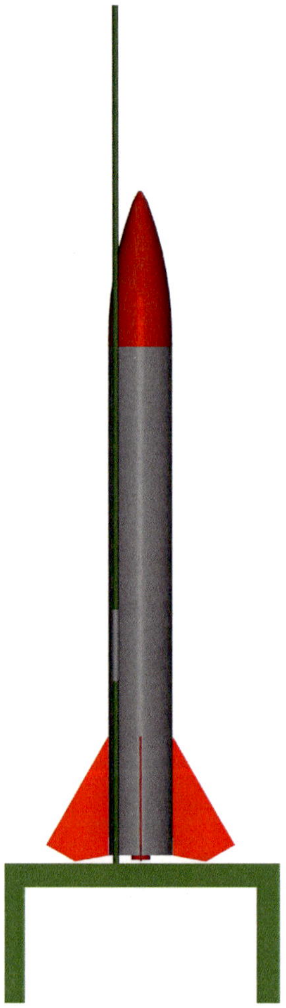

**Figure 4: Rocket at REST**

## 2. PICTURE OF ROCKET AT LIFTOFF

Our next picture is a picture just after liftoff. ***Figure 5*** shows the rocket just after the motor has ignited. Here we see that the force from the motor lifts the rocket up and off of the table.

So what changed? Well two things have changed. The table is no longer acting on the rocket and now that the motor has ignited, **the motor is now acting on the rocket**.

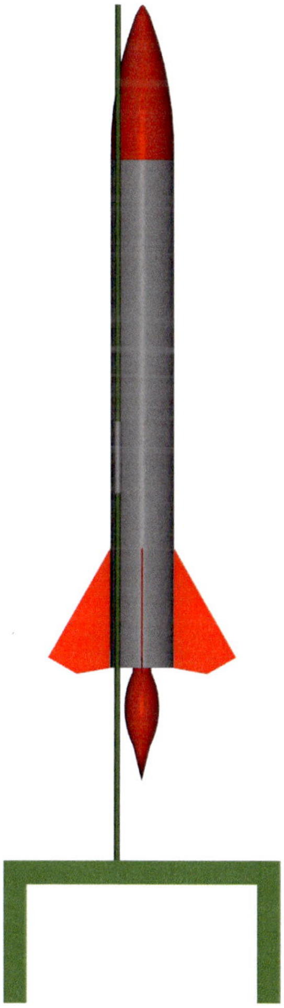

**Figure 5: Rocket at LIFTOFF**

## 3. PICTURE OF ROCKET IN ASCENT

The third picture, ***Figure 6***, represents the rocket during the ASCENT stage. We draw this picture separately because the launch pad (table), is no longer in use, and the motor is accelerating our rocket to the point that now we have to consider the effects of the air on our rocket. So now we have three forces.

One force is due to gravity. Another force is due to the motor thrust. And a third force is a force from the air bombarding the rocket!

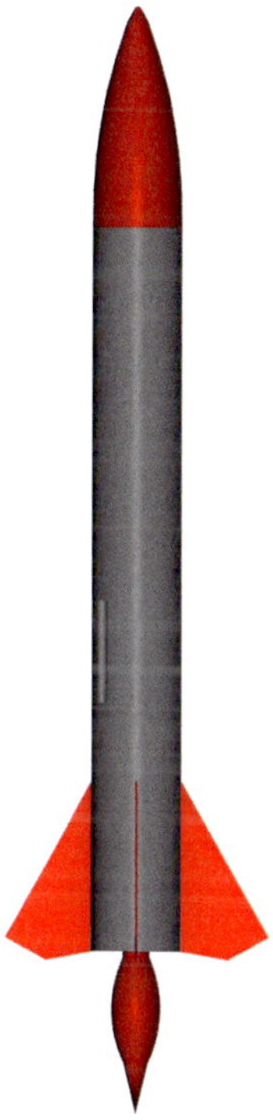

**Figure 6: Rocket During ASCENT**

Now that we have drawn each of the three pictures we can go to our next step of labeling each picture. We can treat each of these as separate problems as we label them and then later, we can combine all three and show graphs of an entire flight!

## 4.2 Label the Picture

1. Draw a picture of the problem
2. **Label the picture**
3. Write a mathematical expression that defines the picture
4. Solve for the unknowns
5. Calculate the results (by hand, by calculator, or by computer application software)
6. Create a lab to verify or deny the work in 1-5
7. Test the results of 1-5 in the lab

Now as we've done in each of the previous labs, we are going to label the picture. This time though, we will label each of the three pictures separately.

## 1. LABELED PICTURE OF ROCKET AT REST

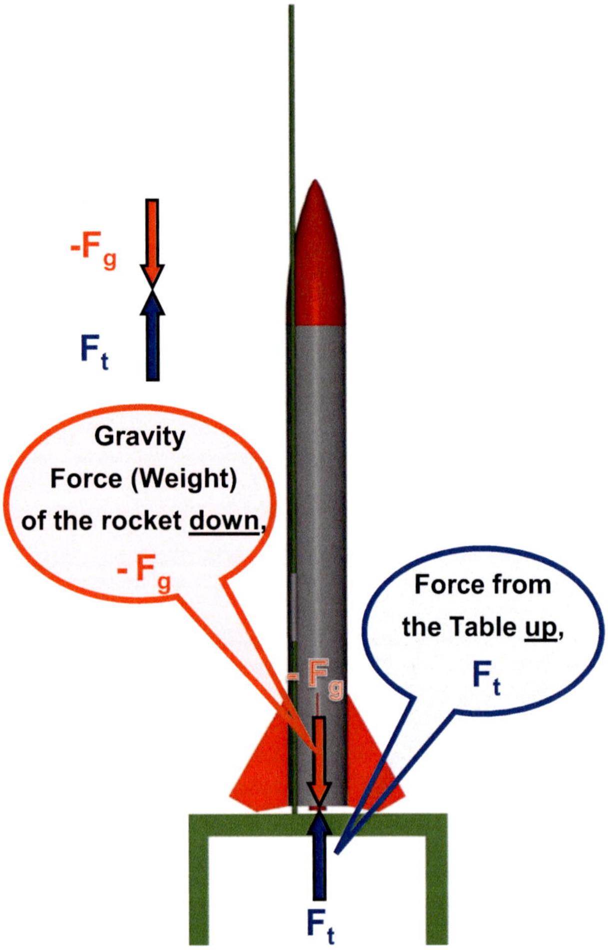


**Figure 7: Rocket at REST**

***Figure 7*** shows our labeled picture of the rocket at **rest**. If you look closely, you can see that there are arrows that are pointed in the direction of the forces. This drawing shows any force in the **downward** position with a negative sign in front of it. We will label each of our three pictures the same way.

Also, another way to illustrate our picture is to just pull the arrows out of the picture to the side so we can concentrate on the labeled arrows. This is called a "**free body diagram**". The arrows are called "**vectors**". So a labeled picture of the problem could just be made from the arrows and the variable names. But for now, we'll draw the picture of the rocket and the vectors.

## 2. LABELED PICTURE OF ROCKET AT LIFTOFF

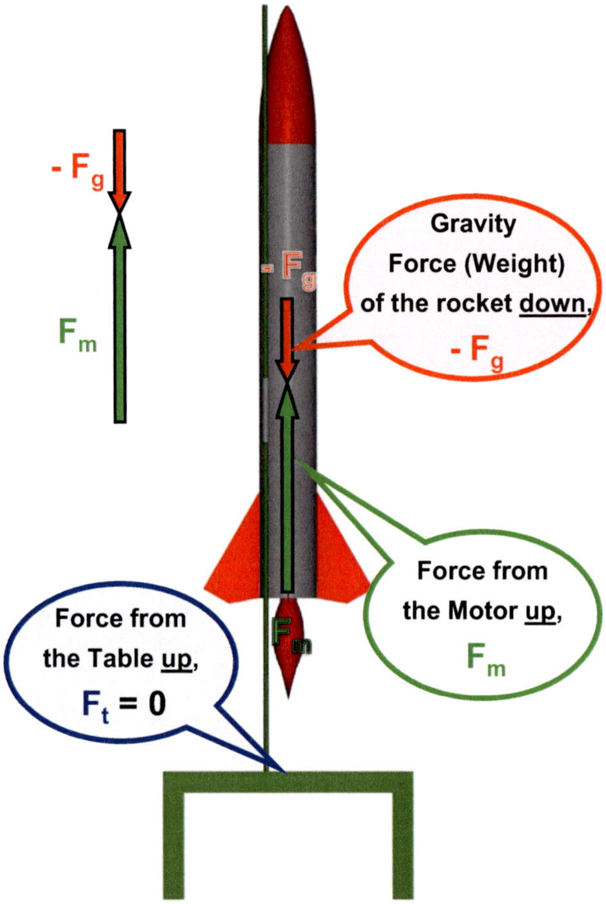


**Figure 8: Rocket at LIFTOFF**

***Figure 8*** shows our labeled picture just after liftoff. Since we have ignited the motor, we have added the force from the motor, $F_m$, in the picture.

Notice that the force of the **motor**, $F_m$, is illustrated so that it is **bigger** than the force of **gravity**, $F_g$ .

Also did you notice that the table force is now zero since it is no longer holding the rocket in place?

Since the motor force has to overcome gravity, it is illustrated as a much larger force than the gravitational force (weight) of the rocket. The gravitational force is again shown as a negative number since the force is in the **downward** position

### 3. LABELED PICTURE OF ROCKET DURING ASCENT

The third picture involves adding another force on our rocket

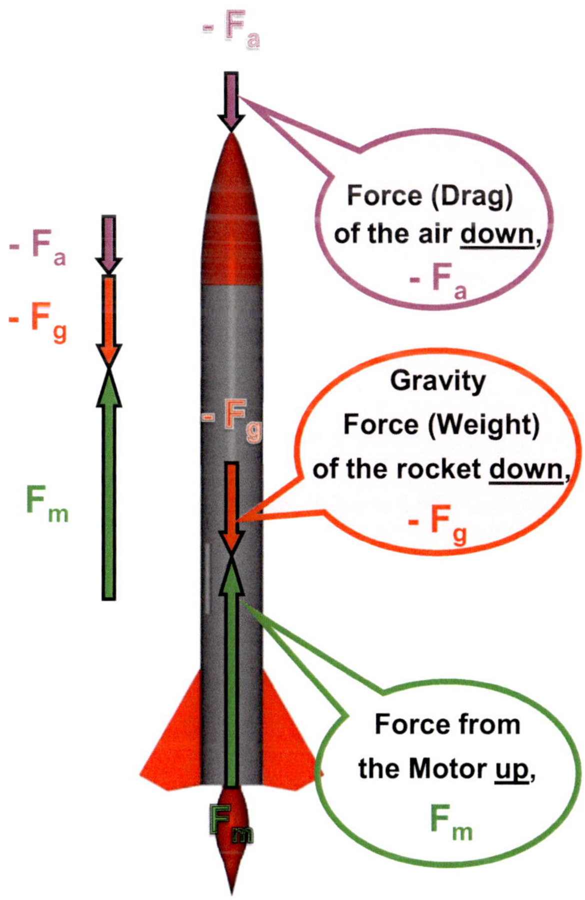


**Figure 9: Rocket During ASCENT**

There's one more force that we need to add to our picture before we can write a mathematical expression that defines the picture.

We've now added a third force to the rocket; another **downward** force. This is the force of the air on the rocket, or force of drag.

We've also eliminated the launch pad since it has no effect on the flight.

Do you remember those annoying little air molecules? They act as a force that keeps us from flying just as the force of gravity, $F_g$.

Even though the force of gravity decreases as the rocket ascends, the force of air, $F_a$, or what we call **drag force** will get bigger as we increase our velocity. If we want to illustrate our picture properly, we will have to account for the forces from the air (atoms and molecules) as well.

Once we label the force due to the air molecules, $F_a$, on our rocket, our labeled picture will be complete and we can move on to our next step in solving a problem.

We'll talk more about $F_a$ in some detail later but for now just remember this is the most challenging aspect to rocket flight.

It's challenging because $F_a$ changes with nosecone design **shapes**, nosecone **surface roughness** and the nosecone **diameter.**

Once in flight, $F_a$ changes dramatically due to the variation in the **velocity of the rocket**. Even air temperature and pressure fluctuations in the atmosphere (**air density**) or lack of an atmosphere affect $F_a$.

**Watch out! The force of air on a rocket cannot be ignored!**

As we did in the previous pictures, we can pull the arrows off of the picture and set them to the side of the rocket. This helps us visualize the directions of the forces in our pictures.

We now have a completely labeled picture. The picture includes the forces due to ***gravity***, the forces due to the ***air molecules***, and the forces of the ***motor***.

Now that we've drawn the pictures for the rocket at rest, the rocket at liftoff, and the rocket during ascent, we're ready to write a mathematical expression for each of the labeled pictures.

**4.3 Write a mathematical expression that defines the picture**

1. Draw a picture of the problem
2. Label the picture
3. **Write a mathematical expression that defines the picture**
4. Solve for the unknowns
5. Calculate the results (by hand, by calculator, or by computer application software)
6. Create a lab to verify or deny the work in 1-5
7. Test the results of 1-5 in the lab

Let's start with the rocket at rest on the launchpad.

## 1. ROCKET AT REST

Our mathematical sentence that describes our picture is:

***The downward force of the rocket on the table is equal to the upward force of the table on the rocket.***

Our shorthand mathematical expression is:

$$- F_g = F_t$$

Did you notice the minus sign? The minus sign is used to keep track of which way a force is acting.

( ***- minus, down..+ plus, up*** )

If we add $F_g$ to both sides of our symbol sentence, it looks like this:

$$- F_g + F_g = F_t + F_g$$

Now if we look at the left side of the expression, we notice that it is zero.

$$( - F_g + F_g ) = F_t + F_g$$

So if we want to replace the $( - F_g + F_g )$ with a zero (since anything minus itself = 0) our math sentences looks like this:

$$0 = F_t + F_g$$

***"When at rest, the downward force of the rocket on the table plus the upward force of the table on the rocket is equal to zero."***

We can rewrite the expression so the "***0***" is on the right side of our symbol sentence. Now our mathematical expression looks like this:

$$F_t + F_g = 0$$

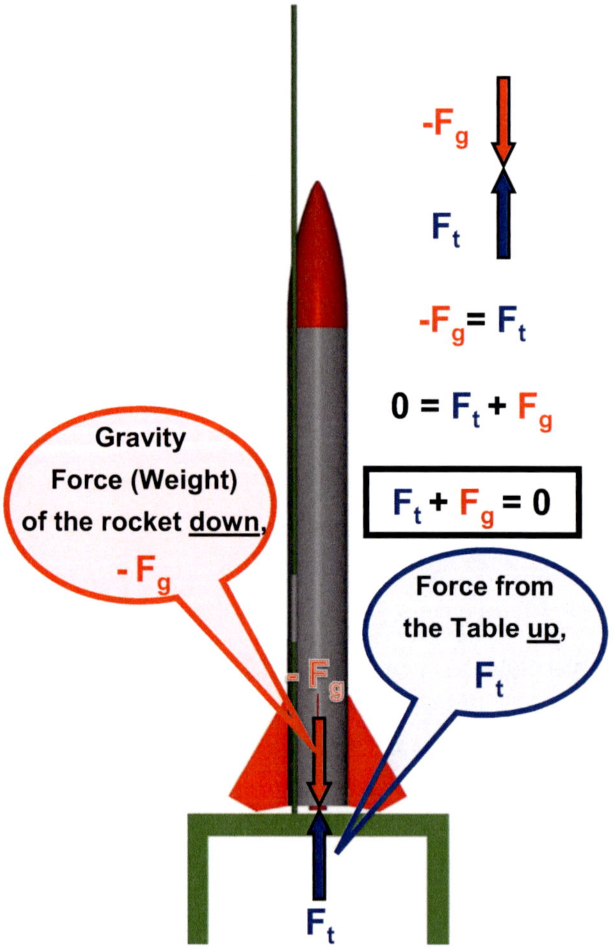


**Figure 10: Rocket at REST**

Any time we add all of the forces from a picture together and they equal **zero**, we can say that the objects are in ***equilibrium***. That is, they are not in motion, but are **at rest relative to each other**. So, if we want our rocket to fly; we have to add a force that will make the rocket move from the equilibrium state to a rocket in motion.

## 2. ROCKET AT LIFTOFF

Once we **ignited the motors**, we **added** the **force of the rocket motor** to our picture. We also showed that the launchpad table is no longer acting on the rocket.

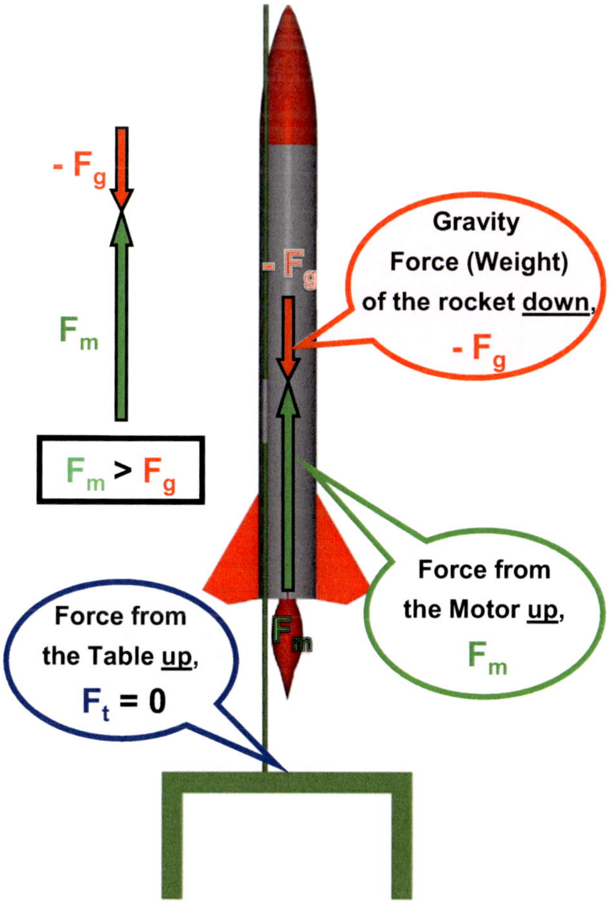


**Figure 11: Rocket at LIFTOFF**

***Figure 11*** illustrates a rocket just as it is taking off from the launch pad. Since the motor force, (see ***Figure 12***) is greater than the gravitational force on the rocket, the rocket will move upward away from the table.

In words we say:

**The force of the motor on the rocket is greater than the force of gravity on the rocket.**

So the mathematical expression that defines this picture is:

$$F_m > F_g$$

**If the motor force was not greater than the force due to gravity, our rocket would not leave the launch pad.**

It would just sit there on the launch pad in a plume of smoke. Even though that looks cool, that's not exactly what we want to happen. Look at ***Figure 11*** and see if it makes sense.

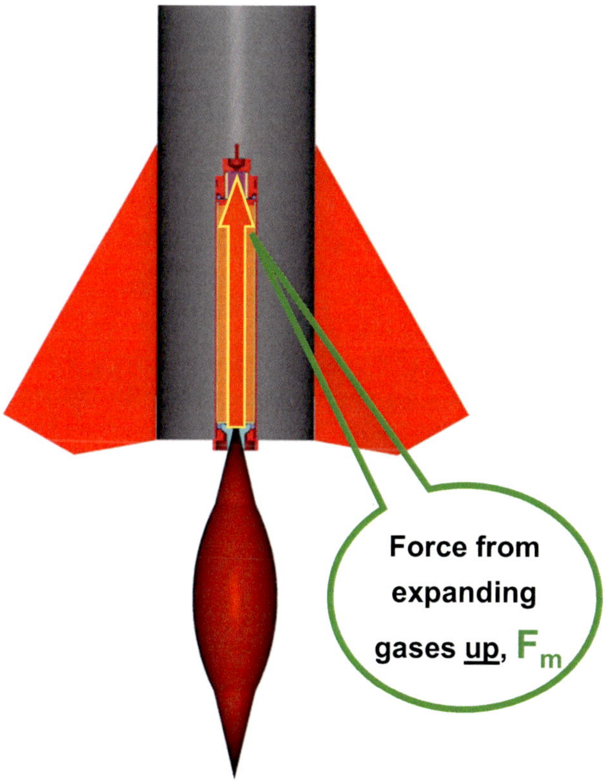


**Figure 12: Motor Forces From Expanding Gases**

## 3. ROCKET DURING ASCENT

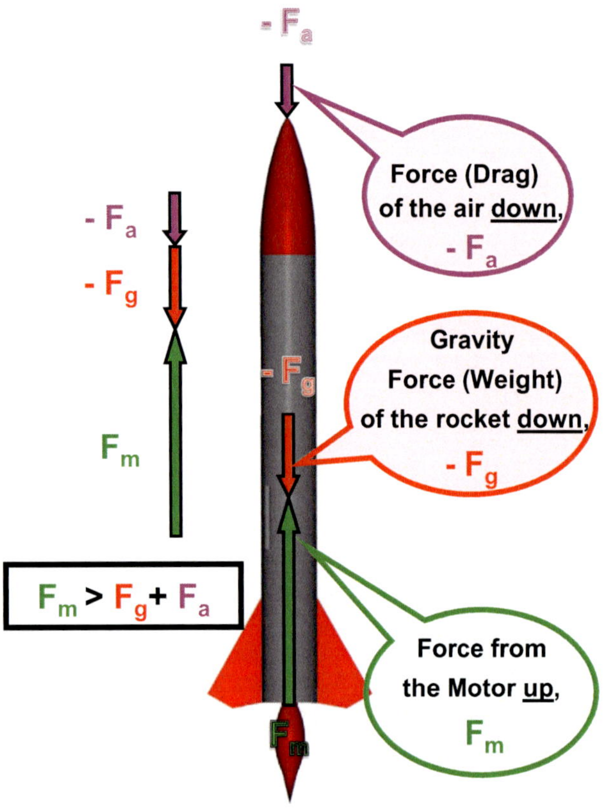


**Figure 13: Rocket During ASCENT**

Look at ***Figure 13***. If you look at the vectors that we drew, you see that the force of the motor has to be greater than the forces of gravity AND the air.

**How much greater does the force of the motor have to be? Well, that all depends on how high the rocket must go.**

The greater the force from the motor, and the longer time the motor burns, determines how high the rocket will fly. We know that the force of the motor has to be greater than the forces of gravity **and** the bombardment of air molecules. Looking at our labeled picture we can say:

***"The force of the motor has to be greater than the sum of the force of gravity and the force of the air molecules bombarding our rocket in flight."***

We can write a mathematical expression that defines the picture in ***Figure 13*** like this:

$$F_m > F_g + F_a$$

Now that we have our **mathematical expression for a rocket in ascent**, we can solve for the unknowns!

*We should remember the considerations for a stable flight we learned in our previous lab. <u>The rocket must be proven stable before the rocket can be flown</u>.*

### 4.4 Solve for the unknowns

1. Draw a picture of the problem
2. Label the picture
3. Write a mathematical expression that defines the picture
4. **Solve for the unknowns**
5. Calculate the results (by hand, by calculator, or by computer application software)
6. Create a lab to verify or deny the work in 1-5
7. Test the results of 1-5 in the lab

The next step is solving for the unknowns. For our mathematical expression, there is really nothing to solve. We have already written it in a useful form. Our mathematical expression that defines the picture in ***Figure 13*** is:

$$\boldsymbol{F_m > F_g + F_a}$$

It says that the force from the motor has to be greater than the sum of the forces due to gravity (the weight of the rocket) and of the air bombarding the rocket in flight.

The force of gravity, $F_g$ is the weight of the rocket in lbf. Suppose the rocket weighs 10 lbf. We can re-write our mathematical expression and put the weight of the rocket in place of $F_g$. When we do that our mathematical expression becomes:

$$F_m > ( 10 + F_a )\ lbf$$

Even though we can get this far in solving our problem, one of the **BIG** unknowns is the force of air, $F_a$ during ascent.

The force of air or drag forces vary throughout the flight. We can solve for $F_a$ by using the following mathematical expression.

$$F_a = \frac{1}{2} * \rho_{air} * v^2 * C_n * S_n$$

In words this mathematical expression says:

**The force of the air on a rocket, is equal to ½ the density of air times the rocket velocity squared times the coefficient of friction of the nosecone times the cross sectional surface area of the nosecone.**

As it turns out, as complicated as this looks and sounds, it's not so bad!

We'll discuss this more in detail later but for now, as you will see later, we can use another "rule of thumb"

Our rule of thumbs says that we can make ***$F_a$ 1 to 1 ½ times the weight of the rocket, $F_g$.***

$$F_a = 1\frac{1}{2} * F_g$$

So if we put this 1 ½ * $F_g$ in our original rocket ascent mathematical expression, it becomes this:

$$F_m > ( F_g + 1½ * F_g )\ lbf$$

$$F_m > ( 2½ * F_g )\ lbf$$

$$F_m > ( 2½ * 10 )\ lbf$$

$$\underline{F_m > 25\ lbf}$$

**4.5** Calculate the results (by hand, by calculator, or by computer application software)

1. Draw a picture of the problem
2. Label the picture
3. Write a mathematical expression that defines the picture
4. Solve for the unknowns
5. **Calculate the results (by hand, by calculator, or by computer application software)**
6. Create a lab to verify or deny the work in 1-5
7. Test the results of 1-5 in the lab

***A motor will supply 4-8 times the weight of the rocket in thrust at liftoff.*** For example, a 10-pound rocket may require 80 pounds of thrust at liftoff. This is to ensure that the velocity of the rocket is fast enough as it leaves the launch rail to force the rocket into a stable straight path. The rocket uses the fins to accomplish this.

The force from the **motor works for** the rocket much in the same way as the force of the **air drag works against** the rocket in flight. Newton's second law comes into play here. He said that if a particle of a certain mass that has a velocity acts on another object, it causes the second object to accelerate based on how quickly the change in velocity is for the two objects.

The motor works by changing a solid or liquid fuel into a gas, thereby increasing the velocity of the fuel molecules. These fast moving gas molecules ***"push"*** the rocket up as they bombard the top of the motor. It does this because the top of the motor is sealed but the bottom of the motor is open. The force going out the open end cannot oppose the force of the gases bombarding the top of the motor.

These motor forces are constantly changing during the flight.

Motors are designed to output a very large thrust force initially, followed by a constant thrust level until; at last, the motor begins to run out of fuel. Once out of fuel, the force due to the motor is zero.

*What about that force on the rocket due to air?* Well as it turns out, it can actually be predicted.

As we said earlier, the forces due to the air, $F_a$, are called **drag** forces.

$$F_a = \frac{1}{2} * \rho_{air} * v^2 * C_n * S_n$$

Let's look at each of these one by one with an example.

$$F_a = \frac{1}{2} * \rho_{air} * v^2 * C_n * S_n$$

- The largest cross sectional *surface area* of the rocket that "slices" through the air, $S_n$, much like your open hand or fist would when you place it outside of your car window.

If you have ever held your hand out of a car window while the car is moving, you probably noticed a dramatic difference in the forces from the air bombarding your hand by either making a fist, or holding your hand open (ball shape to a flat shape). When your hand was in a fist, the forces from the air were much lower than when your hand was held open. This has mostly to do with the shape or CROSS SECTIONAL AREA of your hand when it's open or closed.

When your hand is open the AREA of your hand coming in contact with the air is greater than when your hand was closed.

So when we build rockets the diameter of the rocket and the shape of the nosecone can increase or decrease $S_n$ based on the shapes and diameters that we have chosen to "slice" through the air.

The shape of a nosecone is designed around the velocities the rocket will obtain; less than the speed of sound or greater than the speed of sound.

$$F_a = \frac{1}{2} * \rho_{air} * v^2 * C_n * S_n$$

- The *coefficient of friction*, $C_n$. is a measurement of surface resistance to motion or roughness of the nosecone. The smoother the surface of the nosecone the lower the coefficient of friction.

If you paced a nylon glove on your hand, before you put your hand out the window, the coefficient of friction for the glove would be less than the coefficient of your hand and would lower the value of $C_n$.

$$F_a = \frac{1}{2} * \rho_{air} * v^2 * C_n * S_n$$

- The *velocity*, ***v***, of the rocket is a measurement of Length per unit of Time. We can use units of feet per second, ***ft/s*** or miles per hour, ***mph.*** We could even use furlongs per fortnight, ***fpf***, too!

If you ride in a car and try to hold your hand steady while placing it outside a car window, you will notice it is much harder to hold it steady as the car moves faster

This is the part of the $F_a$ that changes dramatically during the ascent stage of a rocket.

**Notice that the velocity is SQUARED. It's written like this:**

***velocity * velocity, (v*v), $v^2$***

In other words, if your rocket velocity increased from 1 mile per hour, (mph) to 20 miles per hour (mph), and you had an initial force of 1 pound on your rocket due to the air molecules, the force due to the air on the rocket when its traveling 20 times faster would be **400 times bigger.**

mph = miles per hour = miles/hour

At one mile per hour:

$v_1^2$ = ($v_1$ x $v_1$) = 1 mph x 1 mph = **1 mph$^2$**

(1 x 1) = 1 and (mph x mph) = mph$^2$

At twenty miles per hour:

$v_{20}^2$= ($v_{20}$ x $v_{20}$) = 20 mph x 20 mph = **400 mph$^2$ !**

The units for $v^2$ are miles$^2$ per hour$^2$

OR: miles$^2$/hours$^2$ OR mph$^2$

**So just increasing our speeds by 20 makes our $F_a$ 400 times bigger!**

$$F_a = \frac{1}{2} * \rho_{air} * v^2 * C_n * S_n$$

- The *density of air*, $\rho$, really means "how many air molecules are present for a particular day for a particular volume."

Suppose you took a jar outside and collected a sample of air for a couple of days. (Don't forget to put a lid on your jar each day after you collect your sample).

As you counted the molecules each day, the number of molecules would be different depending on the temperature and the barometric pressure the day you took the sample. If you knew how much one molecule weighed, and if you could count all of the molecules, you could multiply the weight of each molecule by the number of molecules you collected to get a total weight of your sample. Each day you would probably have a different count. If the number of molecules were different, the weight of the sample of molecules would also be different each day.

The number of molecules/jar is important since the rocket has to push its way through the air. More molecules/jar means higher drag forces!

**As the rocket velocities increase, what force acting against the rocket increases?**

**How can you make the shape of a rocket lessen the effects of the air resisting the rocket's motion?**

We have mathematical expressions that define all of these changes during the flight, and we can simulate what a flight will actually be like from the time a rocket lifts off, until the time it lands on the ground.

What's important here is that we have drawn a picture of a flight and labeled it so we know what forces are acting on our rocket, both on the launch pad and in flight.

Let's look at our mathematical expressions again.

$$F_m > F_g + F_{drag}$$

$$F_m > ( 2\frac{1}{2} * F_g ) \ lbf$$

Remember our expression and rule of thumb? It says that the force of the motor must overcome **at least** 2.5* $F_g$. How much more will depend on how high you need to fly (and how much money you have to pay for fuel).

*For the rocket on the cover; $F_m = 4 * F_g$:*

The following graphs in this section illustrate the mathematical expressions that were created from our pictures. A computer was used to solve the mathematical expressions to predict the flight of the rocket.

It can take a long time to do these calculations by hand so it's handy to have a calculator or even a computer to solve the mathematical expressions.

We are going to look at some graphs that were created using the mathematical expressions from this lab. These graphs include the effects of:

- The weight of the rocket
- The temperature of the air
- The atmospheric pressure
- The density of the air
- The diameter of the rocket
- The coefficient of friction of the rocket
- A specific motor type

These graphs predicts how high the rocket will fly, how fast it will accelerate, what the rocket's velocities are and how long the flight will last. These graphs *do not include the effects of the rocket's **descent** using a parachute or parasail.*

The graphs in ***Figures 14 and 15*** illustrate the differences in the forces on a rocket due to different nosecone shapes. We will compare Rocket #1 with Rocket #2. **The only difference between the rockets will be the shape of the nosecone.**

Look at the graph of the low friction nosecone design, Rocket #1 in ***Figure 14.*** The coefficient of friction is 0.05.

$$C_d = 0.05$$

**$C_d$ in these graphs is like our $C_n$.**

This graph illustrates the forces on the rocket based on its velocity in flight.

When the rocket is at rest, the forces from the air on the rocket are zero.

Looking at ***Figure 14*** again reveals that at 162 ft/s, (110 mph), the rocket reaches its maximum velocity. At that time the rocket now sees an additional ***8 pounds* of force just from the air**!

**That's like pouring a gallon of milk into the rocket payload bay as its ascending into the air.** I suppose that would be a good job for those bees

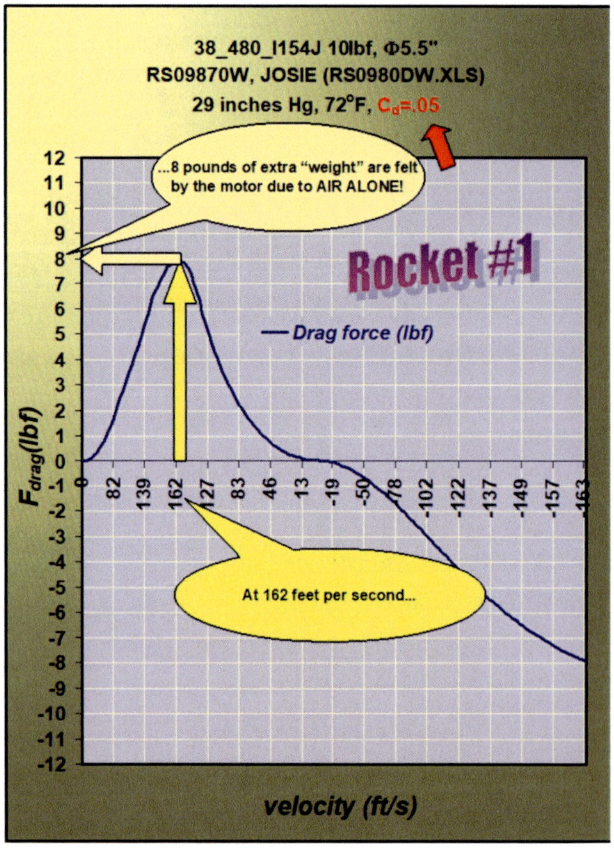


**Figure 14: Force from the air on a rocket in flight with Spherical LOW friction nosecone**

**The drag forces** are almost as much as the force due to gravity (the weight of the rocket)!

This is a real problem for larger rockets. There's a period of time when larger rockets experience what is referred to as "Maximum Dynamic Pressure".

This is a period of time when those pesky molecules and the rocket motor are working against each other the hardest in the flight. Usually, you will throttle down during this period of flight to prevent damage to the rocket. Listen very closely next time you watch a flight. See if you hear the call out "reaching maximum dynamic pressure, throttling down".

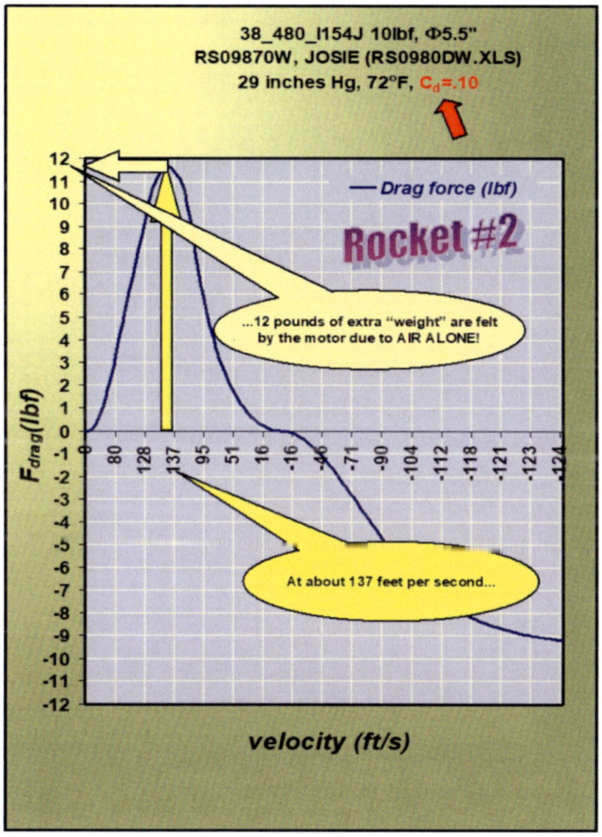


**Figure 15: Force from the air on a rocket in flight with *HIGH friction nosecone***

Now let's compare a rocket that has a **higher friction nosecone** design where the coefficient of friction is **0.10**:

$$C_d = 0.10$$

Let's have a look at the graph of **Rocket #2** in ***Figure 15***.

Again, we see that at rest, the air has no effect on the forces the rocket sees.

But, at 137 ft/s (93 mph), we see that **the higher coefficient of friction causes a drag force that's 50% more than Rocket#1; nearly 12 pounds!**

Now, the force on the rocket, just from a slight nosecone design change, has caused the rocket to "feel" like we added a 1 and ½ gallons of milk to the rocket payload bay during ascent!

The drag forces are now one and a half, (1-1/2), times the amount of forces on the rocket due to gravity!

Another way to look at it is that the rocket now sees 22lbf instead of the mere 10lbf due to gravity.

Even worse, did you notice that we only reached 93 miles an hour? YUCK!!!

As you can see, air molecules cannot be ignored!

We can look at another graph that uses **time on the abscissa** instead of velocity. This graph will show us:

- How high the rocket will fly
- When it will reach apogee (its highest point)
- Its velocity during its flight
- When it will be flying the fastest
- Its maximum acceleration (at launch)

Lets take a closer look.

Remember when comparing these rockets, Rocket #1 and Rocket #2, the ONLY difference between the rockets is the shape of the nosecone.

They both have the same height. They both have the same diameter. They both weigh the same. They both are flying on the same day. They are both flying in the same air (barometric pressure and temperature of the air is the same for both rockets). They both have the same motors.

So for a 10-pound rocket, on this particular day, **ROCKET #1**, ***Figure 16***, will fly to an altitude of **650 feet** in just **7.5 seconds**. It also tells us that the rocket will reach its maximum velocity of 160 ft/s or **110 miles per hour in just 3 seconds!**

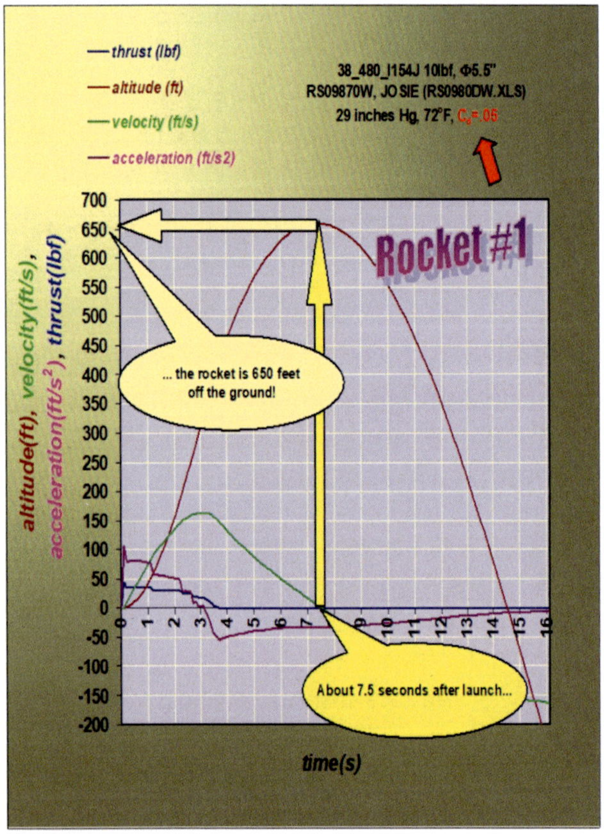


**Figure 16: Flight graph for 10-pound rocket with *LOW friction nosecone***

The second rocket, **ROCKET #2**, ***Figure 17***, reaches its maximum height of 510 ft at 6.5 seconds into flight. It reaches its maximum velocity of 137 ft/s, 93 mph at around 2.75 seconds after liftoff.

**So ROCKET #2 has a shorter flight, and doesn't fly as high, and isn't nearly as fast even though it flew on the exact same day using the exact same motor.**

I know which rocket I want to fly... How about you?

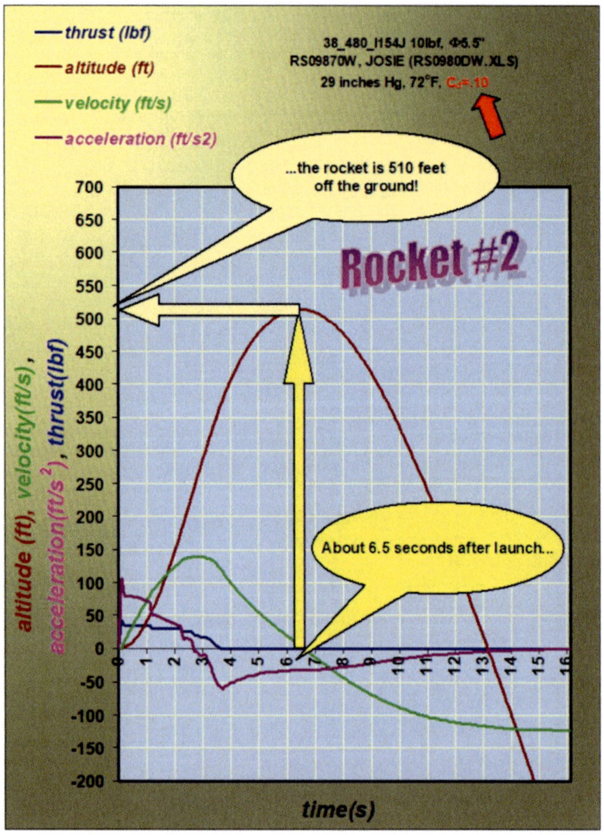


**Figure 17: Flight graph for 10-pound rocket with *HIGH friction nosecone***

**Look at the difference between the forces due to air between Rocket #1 and Rocket #2. Did you notice that the force on the second rocket from the air is now 12 pounds?**

______________________________

______________________________

**Look at the difference in the height and maximum velocities between Rocket #1 and Rocket #2. Can you see any differences?**

______________________________

______________________________

**Why is it important to design rockets, cars, planes or anything else that has to travel through the air with good aerodynamic shapes?**

## 4.6 Create a lab to verify or deny the work in 1-5

1. Draw a picture of the problem
2. Label the picture
3. Write a mathematical expression that defines the picture
4. Solve for the unknowns
5. Calculate the results (by hand, by calculator, or by computer application software)
6. **Create a lab to verify or deny the work in 1-5**
7. Test the results of 1-5 in the lab

We need to see if our mathematical expressions are true by collecting some ***empirical data***. The best and most reliable way is to launch this rocket and see how high it really flies. We can check our mathematical expressions with at least three methods:

1. We can measure the height by sighting the rocket in flight and calculating its maximum height based on the angle we observe at a fixed distance from the launch pad.
2. We could also time the rocket with a stopwatch to see if it reaches its maximum height (apogee) at the same time indicated by our graph.
3. A third method is to review and time the major events with a stopwatch as we watch the on board video, or a ground video of the actual launch.

The third method is my favorite! When we review a video, we can use a stopwatch and time the events as they really happened. Slowing the video down to one-half its speed gives us measurements that are even more accurate. Usually, the measurements come within 10% of our calculated values.

**PROCEDURE FOR TESTING LAB 4:**

You will need some volunteers to help set up the launch site. You should have *one volunteer 1500 feet from the launch pad* with a movie camera and *another volunteer 1500 feet away* from the launch pad to measure an angle from the horizon to the rocket with an instrument that measures altitude, called an *altazimuth*.

Once we arrive at the launch site, we will need to obtain the barometric pressure, temperature and humidity. The values we obtain can be used in a program that solves our mathematical expression. Once we input those values into our program, we will be able to simulate (estimate, predict) how high our rocket will fly.

Collecting the empirical results from our launch will provide us with the <u>*actual*</u> maximum height of our rocket. This measurement will be fairly accurate with only one *angle measurement*. If we want more accuracy, we could add one, or even two more ground to rocket angle measurements to give us a better *actual* maximum height value.

If an on board video camera is used, we will be able to view the video recording from the rocket and we will be able to view the ground video just after the launch. As we review the video, we can use a stopwatch to see if the rocket reaches its maximum height at the same time our program predicted it would.

The angle we obtain from the launch will also give us more information. Our observers will be 1500 feet from the launch site. They will obtain an angle from the ground to the rocket just as the rocket gets to its maximum height and begins to fall back to the ground. When a rocket reaches its maximum height, we call it reaching ***apogee***.

**Here's how we will measure the altitude (how high the rocket flies) based on our angle. *See Figure 18.***

*If you were to draw a line from the launch pad to the observer, and then another line from the observer to the rocket, we would form two sides of a right triangle. The third side of the triangle is formed by the rocket path. If we assume that the rocket flies straight up from the ground at 90°, we can form a three-sided right triangle.*

We will use this right triangle to find out how high the rocket flies. Since we only need two pieces of information from a **right triangle** to solve all the other pieces, we will use the observer's **distance from the launch pad** as one of the pieces of information. The other piece of information will come from the **angle that is obtained** using the *altazimuth* during the launch. We will then solve for the maximum height (altitude) of our rocket.

***Figure 18*** is a picture that illustrates how we will find the maximum height measurement. We can solve for the altitude using a calculator that has *trigonometric* functions, or we can simply use a lookup table.

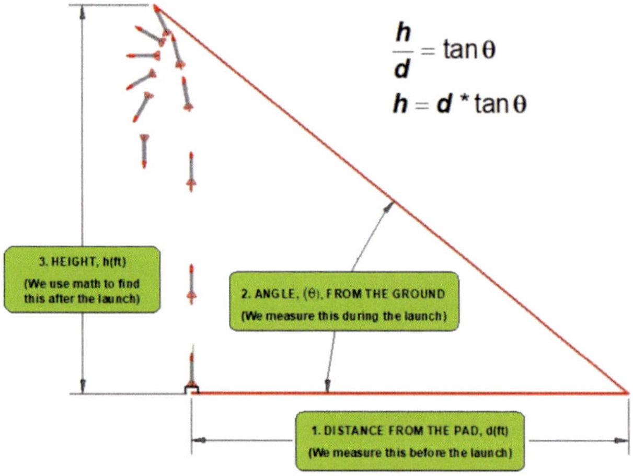


**Figure 18: Determining the rockets maximum altitude**

We use this mathematical expression to solve for the height:

$$\frac{h}{d} = \tan\theta$$

Multiplying both sides by $\frac{d}{1}$ gives us:

$$\frac{d}{1} * \frac{h}{d} = \frac{d}{1} * \tan\theta$$

Since

$$\frac{d}{1} * \frac{h}{d} = h$$

the expression becomes:

$$h = d * \tan\theta$$

This says, "***The maximum height of the rocket is equal to the distance from the launch pad, d, times the tangent of the angle we measured.***"

A table that illustrates ***h*** for an observer measuring various angles from a distance of 1500 feet away from the launch pad is shown in ***Figure 19***.

The graph of the table is also shown in ***Figure 20***.

| **1500** ⇧ | Table for calculating approximate altitude for a rocket using one angle of reference at **1500 feet** from the launch pad. | | |
|---|---|---|---|
| ***d(ft)*** distance from the launch pad (in feet) | **θ** angle measured by observer | ***tan*** **(θ)** tangent of the angle of the observer | ***h(ft)=d(ft)*tan*** **(θ)** altitude of the rocket (in feet) |
| | 0 | 0.00 | **0** |
| | 5 | 0.09 | **131** |
| | 10 | 0.18 | **264** |
| | 15 | 0.27 | **402** |
| | 20 | 0.36 | **546** |
| | 25 | 0.47 | **699** |
| | 30 | 0.58 | **866** |
| | 35 | 0.70 | **1050** |
| **If the observer measured 40°, from a distance of 1500 feet from the launch pad, the rocket's aprroximate altitude was 1259 feet.** ⇨ | 40 | 0.84 | ⇨ **1259** |
| | 45 | 1.00 | **1500** |
| | 50 | 1.19 | **1788** |
| | 55 | 1.43 | **2142** |
| | 60 | 1.73 | **2598** |
| | 65 | 2.14 | **3217** |
| | 70 | 2.75 | **4121** |
| | 75 | 3.73 | **5598** |
| | 80 | 5.67 | **8507** |
| | 85 | 11.43 | **17145** |
| | 89 | 57.29 | **85935** |

**Figure 19: Table of rocket altitudes (1500 foot observation)**

*Look at the table in **Figure 19** and see if you can answer the following question.*

**How high did the rocket fly if an angle of *15°* was observed at *1500* feet from the launch pad? You can use the table in *<u>Figure 19</u>* or the graph in *<u>Figure 20</u>* to find your answer.**

____________________________________

____________________________________

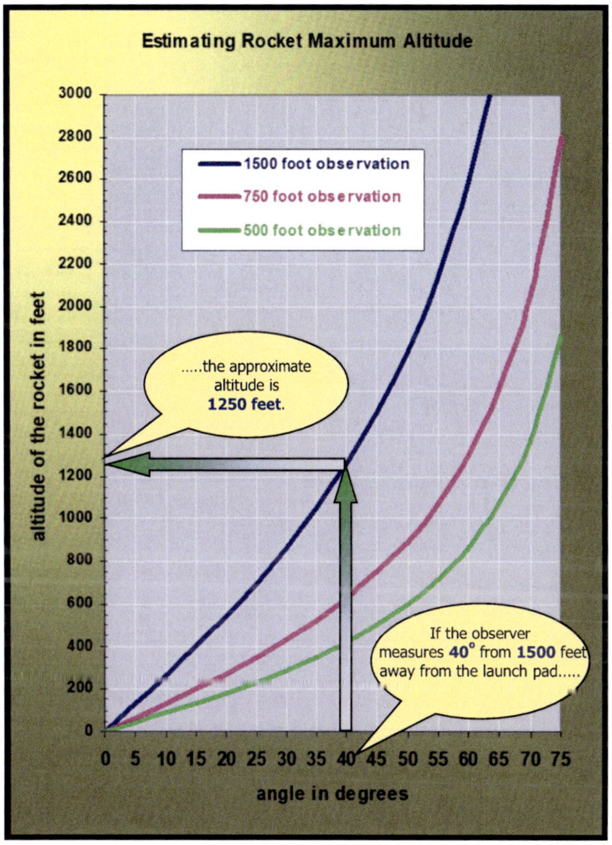


**Figure 20: Graph of rocket altitudes (500, 750 &1500 foot observation)**

**How high did the rocket fly if an angle of *15°* was observed at *750* feet from the launch pad? You can use the table in *<u>Figure 21</u>* or the graph in *<u>Figure 22</u>* to find your answer.**

____________________________________

____________________________________

You can use these same tables and graphs at the launch site to provide you with your altitudes.

| 750 ⇧ | Table for calculating approximate altitude for a rocket using one angle of reference at **750 feet** from the launch pad. | | |
|---|---|---|---|
| ***d(ft)*** distance from the launch pad (in feet) | **θ** angle measured by observer | ***tan(θ)*** tangent of the angle of the observer | ***h(ft)=d(ft)*tan(θ)*** altitude of the rocket (in feet) |
| | 0 | 0.00 | **0** |
| | 5 | 0.09 | **66** |
| | 10 | 0.18 | **132** |
| | 15 | 0.27 | **201** |
| | 20 | 0.36 | **273** |
| | 25 | 0.47 | **350** |
| | 30 | 0.58 | **433** |
| | 35 | 0.70 | **525** |
| **If the observer measured 40°, from a distance of 750 feet from the launch pad, the rocket's aprroximate altitude was 629 feet.** | 40 | 0.84 | **629** |
| | 45 | 1.00 | **750** |
| | 50 | 1.19 | **894** |
| | 55 | 1.43 | **1071** |
| | 60 | 1.73 | **1299** |
| | 65 | 2.14 | **1608** |
| | 70 | 2.75 | **2061** |
| | 75 | 3.73 | **2799** |
| | 80 | 5.67 | **4253** |
| | 85 | 11.43 | **8573** |
| | 89 | 57.29 | **42967** |

**Figure 21: Table of rocket altitudes (750 foot observation)**

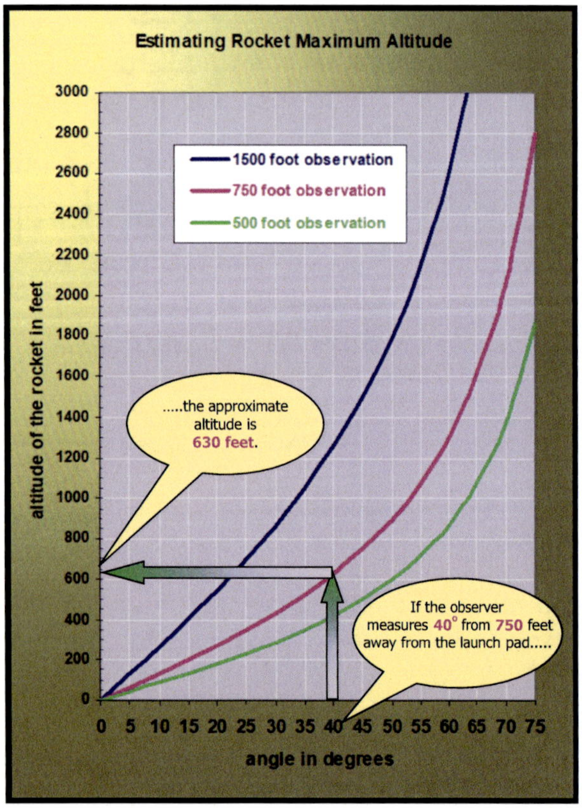


**Figure 22: Graph of rocket altitudes (500, 750 & 1500**

## 4.7 Test the results of 1-5 in the lab

1. Draw a picture of the problem
2. Label the picture
3. Write a mathematical expression that defines the picture
4. Solve for the unknowns
5. Calculate the results (by hand, by calculator, or by computer application software)
6. Create a lab to verify or deny the work in 1-5
7. Test the results of 1-5 in the lab

In the previous lab, we tested a rocket for stability. In this lab, we predicted how high a rocket would fly. In order to check to see if our theoretical work is accurate, we will have to fly our rocket and collect some **empirical results**.

Our ultimate goal is to see whether our altitude and the duration of the flight match our predictions. *The simulations of the flights that are shown in this lab are for an actual rocket. The rocket has been designed to fly to approximately 700 feet.*

While it is in flight, a rocket fitted with a camera or video camera can take pictures and can transmit live video back to the ground by a qualified Technician class Amateur radio license operator .

When you fly your rocket, you can use the charts and graphs that we created to check the actual altitude of the rocket. If all goes well during the flight, we should have some great pictures of the rocket along with some very nice video

**Keep in mind that we want to fly our rockets on a nice warm, calm sunny day.**

**Rockets can never be flown if wind velocities reach 20 miles an hour.**

So be patient, wait for good weather and we'll test our math skills as rocket designers.

## 5 Results (Compare math with real life application)

Although we will not be able to see the results of this lab until launch day, we can make some predictions about your flight based on what we have measured. The flight simulations shown predict that a 10-pound rocket will fly to an altitude of approximately 700 feet. This was based on our theoretical results. The math we perform to predict the behavior of a rocket is called the **theoretical results**. We will collect the **empirical results** when we launch our rocket.

Once we measure the altitude of the rocket in flight, we will compare those results with our theoretical results based on our mathematical expression.

The results are the **comparison** of the **theoretical** results with our actual **empirical** results from the launch.

**Theoretical Results: Predicted**

**Empirical Results: Observed**

## 6 Conclusions (What did we learn?)

- In this lab, we created a mathematical expression from a picture to see how much thrust force is required from a motor to make our rocket fly.
- Remember that we created that mathematical expression from a picture.
- We learned that aerodynamic shapes dramatically affect the performance of the flight.
- In this lab, we used an **inequality** in our mathematical expression for our motor force. We learned in a previous lab that a mathematical expression that defines a picture can be an **inequality, <, >, ≥, ≤,** or equality, **=.**
- We learned how to calculate the maximum altitude of a rocket using an altazimuth and a table.

**Can you see why, once we create a mathematical expression, it's very convenient to show it in a table?**

**Can you see why, once we create a table, it's very convenient to show it in a graph?**

Now, as in every lab, go back to the beginning of this lab and review the abstract. Does it make more sense now?

## 7 References and additional reading

**7.1 The Handbook of Model Rocketry by G. Harry Stine**

**7.2 The Parachute Manual by Dan Poynter**

**7.3 National Association of Rocketry, http://www.nar.org**

**7.4 Email: nar-hq@nar.org**

**7.5 http://www.nasa.gov**

**7.6 http://www.Spacex.com**

# 8 INDEX

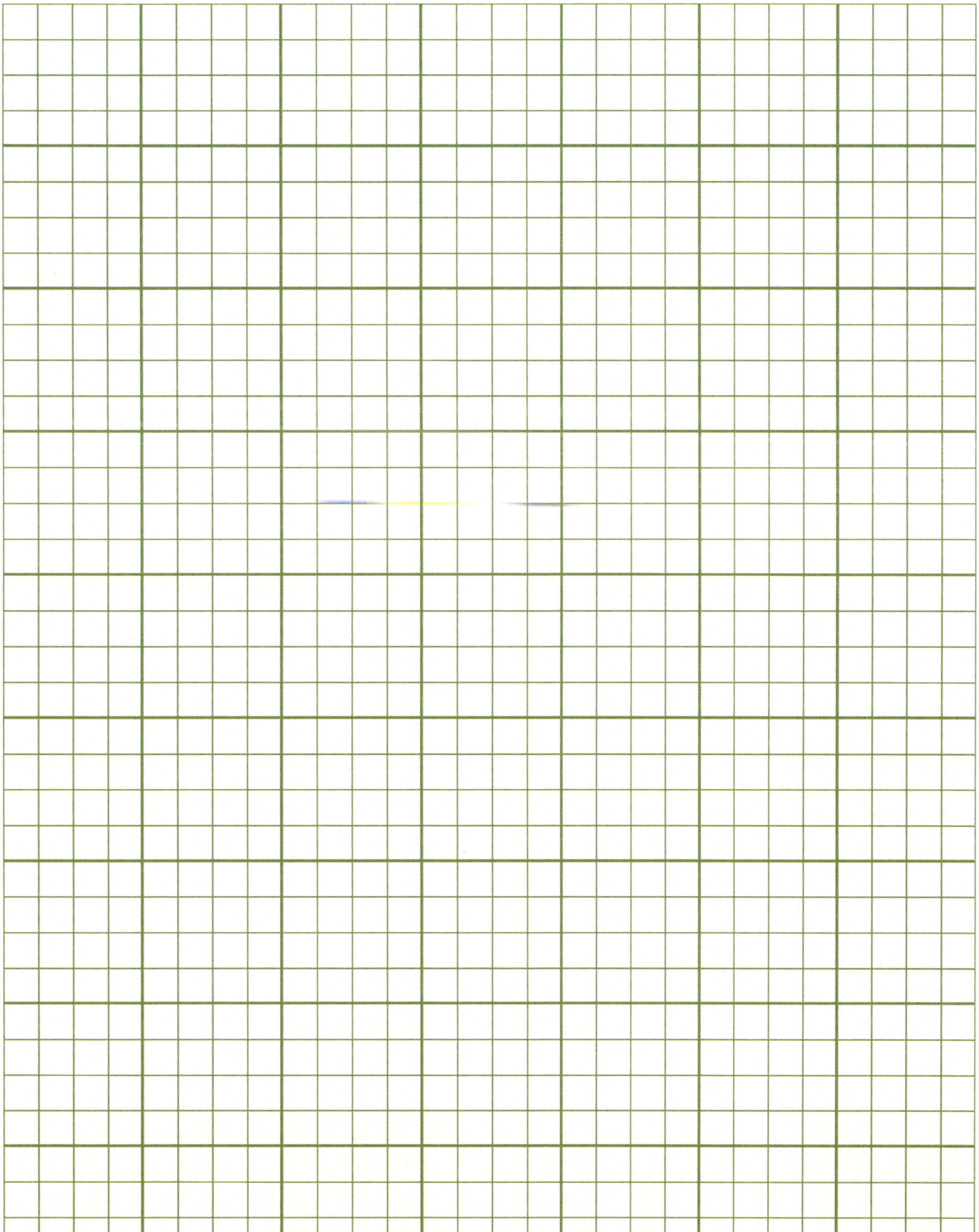

# APPENDIX

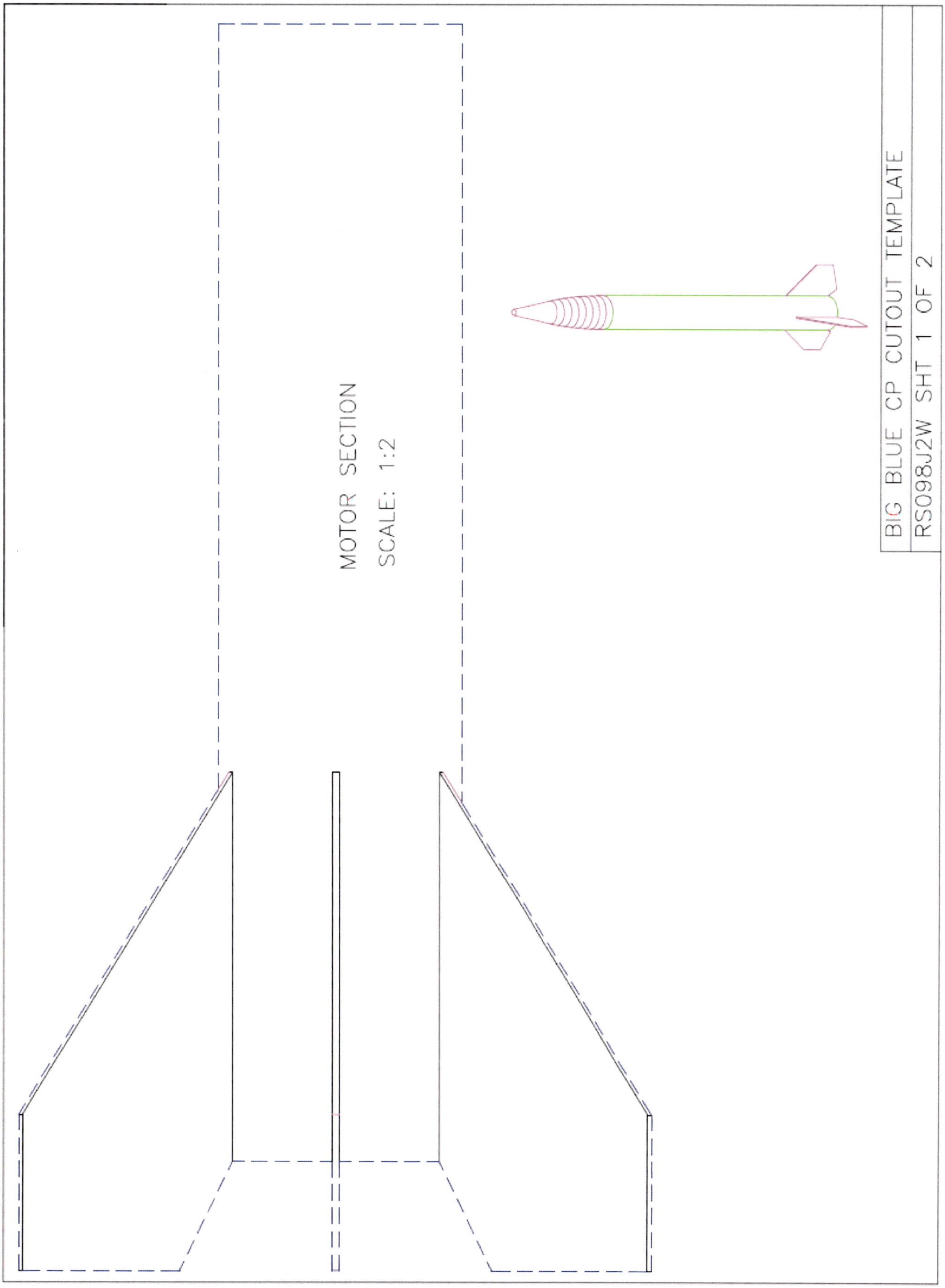
MOTOR SECTION
SCALE: 1:2
BIG BLUE CP CUTOUT TEMPLATE
RS098J2W SHT 1 OF 2

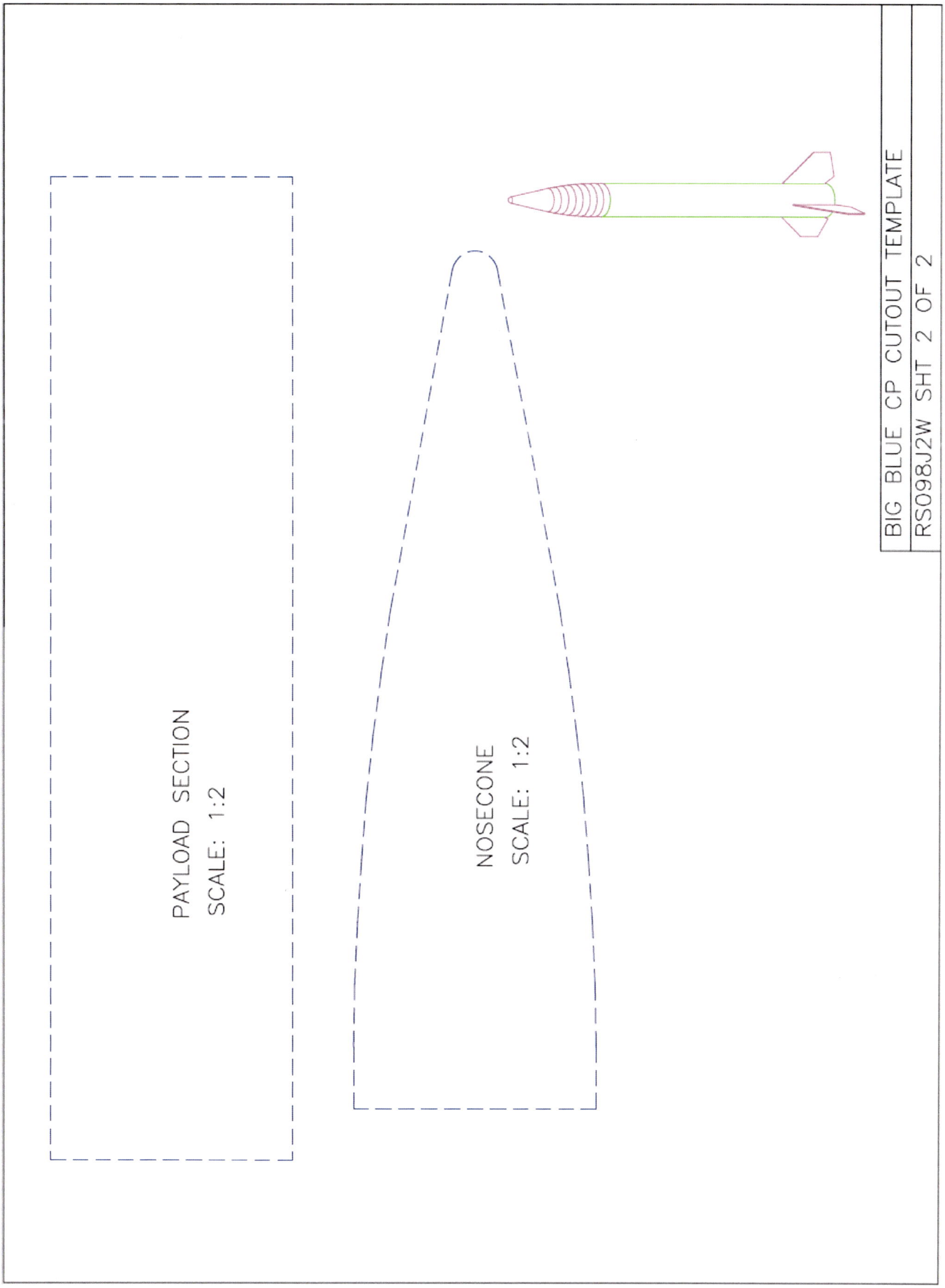
BIG BLUE CP CUTOUT TEMPLATE
RS098J2W SHT 2 OF 2
PAYLOAD SECTION
SCALE: 1:2
NOSECONE
SCALE: 1:2

Made in United States
Orlando, FL
04 November 2026

90201772R00079